THE SERIAL UNIVERSE

Also by J.W. Dunne and available from RESONANCEBookworks

AN EXPERIMENT WITH TIME

NOTHING DIES

www.resonancebookworks.com

FIRST PUBLISHED IN NOVEMBER MCMXXXIV
BY FABER AND FABER LTD.

THIS EDITION PUBLISHED NOVEMBER MMIX
BY RESONANCEBookworks
www.resonancebookworks.com

ISBN 978-0-9559898-6-5

THE
SERIAL UNIVERSE

by
J.W. DUNNE

RESONANCEBookworks 2009
www.resonancebookworks.com

PREFACE

In this book I have tried to give the reader a bird's-eye view of the territory covered by the theory called 'Serialism'. Some of the chapters, greatly condensed, have been delivered in lecture form to the Royal College of Science (Mathematical and Physical Society). But the main outline of the subject is, I believe, clear enough to be appreciated by those who have no special technical knowledge.

Where all is fog, a blind man with a stick is not entirely at a disadvantage. In my case, Fortune presented me with a stick; and I have used this with considerable temerity. Certainly, it has led me somewhere —possibly only into the roadway, where I shall be run over by a motorbus full of scientific critics. But, if I have crossed safely to the other side, then I should like to express my gratitude to Mr J. A. Lauwerys of the University of London, whose continuous encouragement has been the chief factor which has kept me tapping along.

PREFACE TO THE SECOND EDITION

In the first edition of *The Serial Universe,* I used a number of diagrams which had appeared originally in an earlier book of mine called *An Experiment with Time.* In these pictures, the "regress of Time" (which is going to be the subject of our discussion) was exhibited graphically in two dimensions labelled respectively "Time 1" and "Time 2". These were what we may call *"pseudo-Times";* and real Time, not illustrated, was referred to as " Time 3". This, as readers of my latest books, *The New Immortality* and *Nothing Dies,* will realise, was an unnecessary complication: all the arguments could have been greatly simplified had I used only one *pseudo-Time,* Time , leaving real Time, not illustrated, as Time 2. This was done in those chapters of the present book which deal with the supposed wave-particle anomaly. The Relativity chapters, if similarly treated, would have gained greatly; and any persevering reader who finds these difficult is advised to study the easier explanation in Part II of *The New Immortality.*

It is an unfortunate consequence of war that I have been unable to find the time required for the thorough-going simplification of *The Serial Universe* which I intended to effect. I should, for example, have substituted Chapters I to V inclusive, of *Nothing Dies* for Chapters I to XII, inclusive, of the present book, and Chapters XIV and XV of *The New Immortality* for the Chapters XIII, XIV and (in part) XXI which follow here. For my inability to carry this out I can offer only apologies; but these are full and sincere.

I have contrived, however, to simplify slightly here and there, to correct an error in Chapter XXI (corrected already in Chapter XV of *The New Immortality),* and to eliminate a confusion which arose in the presentation of the "Chronaxy" argument on page 111. For directing my attention to this last, I am greatly indebted to Dr. P. J. Strachan, the well-known South African Physiologist.

CONTENTS

Preface		6
Preface to the Second Edition		7
PART I.	THE THEORY OF SERIALISM	
Chap. I.	Meaning of a 'Regress'	14
II.	Artist and Picture	17
III.	Tabular Analysis of a Regress	21
IV.	Regress of Self-consciousness	25
V.	Meaning of 'Observation' in Physics	27
VI.	Regress of a Self-conscious Observer	28
PART II	GENERAL TEST OF THE THEORY	
Chap. VII.	'Now'	34
VIII.	Regress of Time	37
IX.	Regress of 'Reality'. Regress of Physics. Spatial Representation of Time	43
X.	Dimensions, Magnitudes and Mesh-systems	47
XI.	Graphical Analysis of the Time Regress	52
XII.	The Immortal Observer and his Functions	58
PART III	SPECIAL TESTS OF THE THEORY	
Chap. XIII.	An Approach to Relativity	66
XIV.	Velocity of the 'Now'	68
XV.	The Regress in Relativity	75
XVI.	The Physical Outlook of Observer 2	78
XVII.	Quanta, Waves, Particles and the Uncertainty Principle	84
XVIII.	The Regress of Uncertainty	90
XIX.	The Wave Effects	97
XX.	Introducing the Real Observer	102
XXI.	The Place of Brain	106
XXII.	'h'	109
XXIII.	Chronaxy	112
	PART IV	
Conclusion		116
Appendix		117

INTRODUCTION

The men who—little guessing the magnitude of their adventure—set out upon the earliest attempts to understand the world in which we live were rewarded by three surprising discoveries.

They had opened a door—closed till then—in the human mind; and they saw, in a first, dazzling vista, the tremendous powers of abstract reasoning with which Man, all unsuspecting, had been equipped. They had peered behind Nature's mask of happy anarchy; and they stared upon Order—portentous and unassailable. But the strangest discovery was that this orderliness in Nature, and this intelligence in Man, seemed to have been specially created to play partners in a kind of cosmic cotillion of *rationality*. Mind made laws of reason : Nature obeyed them.

They discovered—these early philosophers—that they were wonderful people in a wonderful world. To many, the first of these marvels seemed the more admirable of the two. But there were others of a different temperament. In this respect, indeed, the entire company might have been divided, very early, into two parties. On the one side were those who loved above all things to present abstract problems to that fascinating new toy, the human intellect: on the other were those who found their greatest happiness in the discovery of a new fact to be fitted to facts of nature already ascertained.

Friction between these two divisions must have arisen very soon. For one of the commonest characteristics of a newly-discovered fact is that it appears, at first sight, to be unintelligible. Consequently, every advance of this kind serves to bring into prominence the difference between the pure *'empiricist'* (the man who would put facts before reason) and the pure *'rationalist'* (the man who would put reason before facts). The former is willing to accept the new fact simply because it seems to be a fact : the latter would prefer to withhold recognition until the alleged discovery has proved itself to be reasonable. In the early days of the research, new facts were both plentiful and marvellous; and the cumulative effect of all the little hesitations on the part of the reason-worshippers was, sometimes, considerable. But, always, they caught up again; for the empiricist's structure of facts proved, invariably, in a little while, to be entirely reasonable. Nevertheless, these delays in admitting new discoveries were harmful to the prestige of the rationalists; for every such lagging-behind meant that the empiricists had obtained knowledge (admitted, later on, to be true) which had been established upon a basis other than that of pure reason.

All this, however, was merely first-line skirmishing. In their main position, the rationalists had dug themselves in so deeply that none, save a few complete sceptics, dreamed of trying to dislodge them. Their cardinal tenet—that reason, unaided, could discover the great fundamental truths which facts of experience served merely to illustrate—had been adopted by the metaphysicians as the basis of an energetic inquiry into the constitution of the universe. And the empiricists, although they may have doubted the expediency of the metaphysician's methods, never supposed for one moment that such facts of nature as remained to be discovered would prove to be, at bottom, otherwise than wholly reasonable.

Now, nobody had disputed that reasoning is a machine which deals faithfully with all the material offered to it, provided its owner does not attempt to alter its method of working. But it is a machine which needs feeding with 'premisses', i.e., assertions *presumed* to be true. The rationalists claimed to have discovered the most fundamental premisses of all—basic truths which could not be denied, but which, because they were basic, could not be proved. Knowledge which satisfies that description is said to be 'given', and the supposed given knowledge which the rationalists selected as the base of their edifice consisted of a set of axioms asserting what could or could not exist without self-contradiction. The empiricists, however, were able to point to given knowledge of an apparently different kind. The evidence of the senses is notoriously unreliable, but what none can deny is the existence of *the evidence.* We may doubt what a sensory experience seems to assert; we may be a little vague even regarding the precise character of the experience itself: but we reach, through our senses, a limit to what it is possible for us to deny—we arrive at what is (for us) an undeniable *residuum* which we call the 'sensation', or, in less popular language, the 'sense-datum'.

The fact that the sense-data of the empiricists happened to obey the axioms of the rationalists, and were never self-contradictory, shed no light on the main problem. Was the universe the product of Mind, so that it, and experience of it, must illustrate Mind's axioms? Or did the universe exist independently; and were our infrangible axioms no more, at bottom, than our recognitions of the special kind of *order* which we happened to have discovered pervading that universe, and so, no more than illustrations of our inability to grasp the possibility of any other kind of order?

That question was never answered. An interruption occurred. In the height of the discussions, an Irishman, Bishop Berkeley, threw into the philosophic duck-pond a boulder of such magnitude that the resulting commotion endures in ripples to this very day. He asked an entirely different. question. If sensations such as those of colour, form and feeling, *plus* their derivatives of memory-images, associated 'ideas', concepts and the like, were the sole bases of our knowledge,—the only objects with which we were, or could be, directly acquainted,—what *evidence had we that there existed any substantial, non-mental world at all?*

You may imagine the joyous rallying of rationalists which followed the appearance of this 'Idealism' (as Berkeley's theories were called). No physical universe! Nothing but a vast, collective hallucination! Then Mind was Lord of All.

Philosophy, split horizontally by the division between rationalists and empiricists, was riven vertically by the far fiercer dispute which arose be the idealists and the realists. Peacemakers suggested an 'intuitive' knowledge of objective reality. Voluntarists argued that this intuitive knowledge was knowledge of opposition to 'Will'.

But the rationalists wished to limit the intuitive bases of their structure to cognition of the three 'Laws of Thought'; while intuition, if it existed, would be a process beyond reach of the empiricist's tests. But the idealists were not only assailed from without: they were betrayed from within. There arose very quickly a critic who said, in effect, 'What is all this talk about a "collective" hallucination? *If* all that I can know directly are my sensations, and no external universe can be inferred from these; then I have no reason to suppose that there exists any mind other than my own. I am the only experient, and the hallucinatory world is my world, and mine alone.' The logic of the argument seemed to be unassailable. No answer could be found then: none was found later.

Most of the idealists were unable to face the unescapable consequence of their thesis. 'Solipsism' (as this completed theory was called) proved too indigestible for any but the absolute purists. The rationalist quarter, moreover, had been worried considerably by the logical discoveries of Hume, who proved that, if the world of sense-data were all that existed, a Mind controlling this display would be as hallucinatory as an external world. In the end, so far as the majorities were concerned, the rationalists abandoned their rationalism, the empiricists discarded their empiricism, and both agreed to accept the external world as 'given' by some concealed process which (it was hoped) would prove some day to be both rational and empirical, but which, till then, could not be classified as anything beyond that irrational and intangible thing—*intuition*. And so, on a basis of intuition, Science came into its own.

Progress was now rapid. Rationalists and empiricists hurried hand in hand towards a goal which showed ever clearer and more brilliant. It was discovered, with profound relief, that the real universe consisted of conglomerations of little round things like billiard balls, called 'atoms'. Electricity was found to be a modification of an all-pervading elastic solid called 'aether'. There were laggards who pointed out that the primary sense-*data*—such as colour—could not be composed of, or accounted for by, either billiard balls or waves; but the gleam of the *Absolutely Reasonable* shining just ahead blinded nearly all to the mists of irrationality gathering on either side. They reached that gleam and it vanished at that moment. The solid atoms fled away. In their places lay voids tenanted by minute specks too unreal to possess both precise position and precise velocity. Did I say 'specks'? They were not specks, but waves filling all space. Photographs proved it. Worse, each of these wave-entities needed a whole three-dimensional world to itself, so that no two could be together in the same ordinary space. Did I say 'waves'? I am sorry, they were specks in one and the same space. Experiments proved it, and they could be even counted by a specially designed apparatus. They were not mixtures of specks and waves : each was, definitely, both. A strange phantasmagoria. It was founded upon the indubitable existence of a tiny, irreducible, four-dimensional magnitude called the 'Quantum' —itself the very acme of irrationality. And the behaviour of this irrational universe could be calculated only by the aid of a specially invented 'irrational' algebra.

On another side they were faced by the world of Relativity. Here the aether had either disappeared, or it survived merely as a purely personal appendage—as subjective as any Solipsist could desire. Space and time had not vanished : they had done worse : they had become interchangeable. And the 'space-time' world of the relativists appeared to be governed throughout its expanse by the square root of minus one—famous in mathematics as the basic 'imaginary' number.

Now, reasoning must start from 'given' knowledge, and that knowledge is, consequently, not rational. No science, therefore, proposes to explain, or expects to explain, the existence of whatever it accepts as the fundamental realities. But its object is to employ those elementary indefinables as characters in a narrative of rational happenings. And there is a fairly general feeling that, in the tale which our science offers us to-day, the irrationalities are far too numerous. It is a true story; but it looks as if, somewhere, somehow, it had been made into 'printer's pie'. The right words are there, but they seem to be in the wrong places; and there is more than a suggestion that paragraphs which ought to have been consecutive have become superimposed. Waves, particles, space-time, quanta and even sense-*data* must, we feel, fit together in some simpler fashion. And we suspect that, if only we could discover that scheme, all these surplus irrationalities would vanish, leaving us with nothing that was not obvious and expectable to the most ordinary intelligence, and with nothing more obstreperous than the two basic indefinables of Mind and Matter.

THE SERIAL UNIVERSE

PART I

THE THEORY OF
SERIALISM

CHAPTER I

MEANING OF A 'REGRESS'

A 'series' is a collection of items linked together, chain-fashion, by some recurrent relation. The notion of series has reference, always, to some underlying unity; this is implicit in the fact that the separated items are related to one another.

The distinctive items of a series are called its 'terms'. For example, if we regard a child as a creature who had a parent who had a parent who had a parent, etc., etc.; the child is the first term, his parent the second term, and his grandparent the third term of a receding series. And, if we tabulate that series thus:

1st term	2nd term	3rd term	4th term	
A child of	a parent who was child of	a parent who was child of	a parent who was child of	etc., etc.,

the relation between the terms becomes readily apparent.

We know, from various biological indications, that this particular sequence stretches back to before the dawn of history. But the old-time philosophers thought that it must either recede to a time infinitely remote, or have been started by some magical act of creation. And it is rather interesting to consider what were their grounds for that assumption.

If we look at the first term in the table, we find there an individual to whom we have allotted only one character—the character of being a child. Now the fact that every child has or had a parent is merely a truism; it is asserted already in the meaning attaching to the word 'child'. And, taken by itself, it does not compel us to entertain the notion of remoter ancestors. But suppose we go on to the second term. We come to a person who is declared to possess a *double* character—a person who is both parent and child. As a parent, he is related to the first-term individual already examined; and, as a child, he must be related to some ancestor not yet taken into account. Now, the early philosophers supposed, wrongly, that it was a matter of logical necessity for *every parent to be also a child.* If that had been true, the series, obviously, would have been bound to extend backward to infinity.

The point—the point which is so often overlooked—is this : The extension of a simple series to infinity involves some necessarily dual character in its terms. But, to discover that dual character, we must trace the series as far as its *second* term. A study of the first term (such as the child in the above example) with its single character, will yield us only half the required information. And it may be noted that the third and remaining terms do no more than repeat the information already asserted by the second term. All the remoter individuals in the purely imaginary example we have taken would have possessed the double character of being both parent and child; but we could have discovered that from an examination of the second term alone.

In brief: *Every simple series to infinity is the expression of some logical fact which is asserted in the second term but not in the first.*

And, as we shall see later, it may be impossible to exaggerate the importance, to the human race, of this very simple characteristic of a simple infinite series.

Now, a series may be brought to light as the result of a question. Someone might enquire, 'What was the origin of this man?', or a child learning arithmetic might set to work to discover what is the largest possible whole number. The answer to the first question has not yet been ascertained : the answer to the second can never be given. It will be seen, however, that the reply in each case must develop as a series of answers to a series of questions. In the first instance, we reply that the man is descended from his father; but that only raises the further and similar question, 'What was the origin of his father?'. In the second case, the child will discover that 2 is a greater number than 1; but he is compelled to consider then whether there is not a number greater than 2—and so on to infinity. A question which can be answered only at the cost of asking another and similar question in this annoying fashion was called, by the early philosophers, 'regressive', and the majority of them regarded such a 'regress to infinity' with absolute abhorrence.

Their attitude is easy to understand. They wished to regard the universe as something completely explicable. To admit that there were questions with answers which receded as a rainbow recedes, was, in their opinion, to admit, before they started, that their task of explaining everything was fore-doomed to failure. Then, again, a considerable number of the early philosophers supposed that the universe must be, at bottom, something extremely, even childishly, simple; a naive theory which involved that to every question there must be a simple and straightforward answer. This provided another reason for the ancient dislike of regressions. And we must add to the list that very numerous class which wished, and still wishes, from motives of policy, to divide the world sharply into things which are comprehensible and things which are incomprehensible. To such persons, a question which is answered by an 'infinite regress' is *anathema,* because it provides, very obviously, a class between the two divisions.

In brief, it was universally recognised that a regress might be logically incontrovertible; men moulded their lives and their sciences upon the immense stock of reliable information provided by the study of these incompleted series of questions and answers; and yet the regress to infinity was looked upon as being, in some fashion apparent only to intuition, not actually untrue, but not precisely that aspect of the truth which it was the business of philosophy to discover.

They were quite unable to put this feeling into words. They wandered off into loose talk of 'complexities', which was a dubious charge, and of 'contradictions', which was a libel unjustified in anyone with any pretensions to intelligence—for a contradiction produces no regress at all, and the whole trouble about the infinite regress is its damnable logicality. If the truth of the premiss (i.e., the double character of the second term) is acknowledged, the regress becomes mathematically inevitable. Yet the feeling has persisted to this day : it crops up afresh whenever some new regression, to the sight of which we have not grown accustomed, is discovered. And Bradley, perhaps, gave it its nearest approach to verbal expression when he said, 'Reality cannot be an infinite regress'.

The answer, I think, is this :

The truth or falsity of Bradley's dictum depends upon the meaning it attaches to the word 'reality'. If it refers to reality pure and undefiled by any attempt at translation into terms of human comprehension, his statement, probably, is true (though you must not ask me to give reasons for that belief). But if the word means reality in the scientific sense,— rational *cum* empirical reality,—then the assertion is, definitely, wrong. The difference is that which lies between 'things as they are' and 'things as they seem to be'. Of 'things as they are' we know nothing rational; and, if we suspect Bradley to be right, it is merely because of the feeling of dissatisfaction aroused in us by any regress. But of 'things as they seem to be'—things as they affect an *observer*—we can say a great deal. As I hope to show in this book, we can say, with absolute assurance, that 'reality' *as it appears to human science* must needs be an infinite regress. And it is only when it is expressed in that form that we can treat it as the reality upon which we can rely.

CHAPTER II

ARTIST AND PICTURE

A certain artist, having escaped from the lunatic asylum in which, rightly or wrongly, he had been confined, purchased the materials of his craft and set to work to make a complete picture of the universe.

He began by drawing, in the centre of a huge canvas, a very small but very finely executed representation of the landscape as he saw it. The result (except for the execution) was like the sketch labelled X_1 in FIGURE 1.

FIGURE I.

On examining this, however, he was not satisfied. Something was missing. And, after a moment's reflection, he realised what that something was. *He* was part of the universe, and this fact had not yet been indicated. So the question arose : How was he to add to the picture a representation of himself?

Now, this artist may have been insane, but he was not mad enough to imagine that he could paint himself as standing in the ground which he had already portrayed as lying in front of him. So he shifted his easel a little way back, engaged a passing yokel to stand as a model, and enlarged his picture into the sketch shown as X_2 (FIGURE 2).

FIGURE 2.

But still he was dissatisfied. With the remorseless logic of a lunatic (or genius—you may take your choice) he argued thus:

17

This picture is perfectly correct as far as it goes. X_2 represents the real world as I—the real artist—suppose it to be, and X_1 represents that world as an artist who was *unaware of his own existence* would suppose it to be. No fault can be found in the pictured world X_2 or in the pictured artist, or in that pictured artist's picture X_1. But I—the real artist—am aware of my own existence, and am trying to portray myself as part of the real world. The pictured artist is, thus, an incomplete description of me, and of my relation to the universe.

FIGURE 3.

So saying, he shifted his easel again, seized his brush and palate, and, with a few masterly strokes, expanded his picture into X_3 (FIGURE 3).

Of course, he was still dissatisfied. The artist pictured in X_3 is shown as an artist who, though aware of something which he calls himself, and which he portrays in X_2, is not possessed of *the knowledge which would enable him to realise the necessity of painting X_3* the knowledge which is troubling the real artist. He does not know, as the real artist knows, that he is self-conscious, and, consequently, he pictures himself, in X_2, as a gentleman unaware of his own existence in the universe.

The interpretation of this parable is sufficiently obvious. The artist is trying to describe in his picture a creature equipped with all the knowledge which he himself possesses, symbolising that knowledge by the picture which the pictured creature would draw. And it becomes abundantly evident that the knowledge thus pictured must always be less than the knowledge employed in making the picture. In other words, *the mind which any human science can describe can never be an adequate representation of the mind which can make that science.* And the process of correcting that inadequacy must follow the serial steps of an infinite regress.

This pictorial symbol does not lend itself very readily to detailed analysis, and we shall make little further use of it. It provides, however, an excellent illustration of the differences which underlay the views of (1) the old-fashioned man of science, (2) the materialist, and (3) the average philosopher. The classical physicist held (wrongly, as we shall see) that the picture X_1, which contains no reference to an artist, ought to prove self-consistent and self-sufficient. The materialist held (wrongly, as *we* have seen) that the second picture, X_2 (q.v.), would describe closely enough for practical purposes the relation between man and his universe. He omitted to note that the artist shown in that picture is only the first term of a regressive conception, and that, to get at the practical information which is expressed in such a series, we must study the *second-term* individual. The average philosopher found himself in a quandary. He could see that the materialist was at fault, but he was unable to point to the error without pointing to a regress which he did not know how to handle. Consequently, he hesitated—while the error gained adherents. And thus there became established that picture, so popular to-day, which exhibits the universe as nothing more or less than an indifferently gilded execution chamber, replenished continually with new victims. The materialist was scarcely to blame : he was honestly myopic. But the philosopher was a politician.

The regressive picture of our symbol contains, not only a series of artists of increasing capacity, but also a series of the landscapes which such imagined individuals would draw. One might suspect that the details of those landscapes—the hills and trees and houses—ought to bear some witness to the increasing skill of the draughtsmen and exhibit a serial progress towards a regressive perfection. Now, we shall discover, in the course of this book, that the entire symbol, with this additional interpretation, is absolutely correct. This means that, whatever the universe may 'be' in itself, all *sciences* thereof must be regressive, so that we are faced with what is, for all empirical purposes, a serial world. And, when we recall that the relation of such a world to ourselves—the repetitive relation which makes the regress—is given by the second term and not by the first, it will become evident that the theory of the 'execution chamber' was a particularly ludicrous blunder.

Omitting the arguments, the conclusions of the theory I call 'Serialism' are, briefly, as follows.

We are self-conscious creatures aware of something which we are able to regard as other than ourselves. That is a condition of affairs which it is impossible to treat as rational (i.e., systematic) except by exhibiting it in the form of an infinite regress. Consequently, the first essential for any science which can satisfy us as fitting the facts of experience is that it shall employ some method of description which is suitably *regressive*. It turns out that the possibility of viewing all experience in terms of 'time' provides us with just the method of description required. The notion of absolute time is a pure regress. Its employment results in exhibiting us as self-conscious observers. It introduces the notion of 'change', allotting to us the ability to initiate changes in a change-resisting 'not-self'. It treats the self-conscious observer as regressive, and it describes the external world as it would appear to such a regressive individual. Thus it fulfils all the requirements of the situation. But time does more than that. By conferring on the observer the ability to interfere with what he observes and to watch the subsequent results, it introduces the possibility of *experimental* science. The notion of experiment implies always an interference with the observed system by an observer outside that system. This is the cardinal method of physics, which postulates, thus, from the outset the possibility of interference with *every* system by an observer who, in relation to that system, is 'free'. The essential point here, however, is that physics, as a science of experiment,—of 'alter it and see',—is based upon the notion of time. So, for that matter, are all our systems of practical politics, ethical or otherwise. In that way only—by the employment of this flagrantly regressive method of description—have we been able to convert our otherwise irrational knowledge into a systematic and serviceable scheme.

But is this regressive way the *proper* way to describe the universe? That question has little, if any, meaning. Is 'decimal point three recurring' the 'proper' way to describe 'one-third'? The regress of the recurring decimal and the regress of time both rank as series to infinity; and, though the former series is 'convergent' and the latter 'parallel', the underlying principle in each is the same. There is probably another way of describing the universe, just as there is another way of describing one-third. We use the decimal method because it is convenient for our purpose and just as valid as the other. We use the time regress because it gives us a valid account of the universe *in its relation to ourselves,* that is, in its reaction to experiment. It is the proper method for-its purpose, and I know of no profounder meaning in the word 'proper'. But this I do know : It is impossible to imagine a more effective way of *losing* knowledge than that of expressing it in the form of an infinite regress and then restricting attention to the first term alone. And that is what mankind has been doing.

All talk about 'death' or 'immortality' has reference to *time,* and is meaningless in any other connection. But a time-system is a regressive system, and it is only in the lop-sided first term of that regress that death makes its appearance. It will become clear in the course of this book that, in second-term time (which gives the key to the whole series) we individuals have curious - very curious - beginnings, but no ends. Is that a horrible thought? Perhaps. But I do not think so. The present-day terror of immortality is based, almost entirely, upon an imperfect appreciation of what that immortality means. We try to imagine it as fitted somehow into the first-term world, (where, of course, it won't go), and so plague ourselves with a lugubrious picture of bored individuals dragging memory's ever-lengthening chains, desperately sick of themselves and the world and all that therein is, craving an extinction which they cannot find. We imagine, in fact, our present kind of daily life continued for ever. If that were true, there could be no act more cruel than the act of giving birth to a child. But, fortunately, our immortality is in multi-dimensional time, and is of a very different character.

And now for the proofs. These must develop, so to say, backward. We must take the world of our present-day knowledge, show that it is regressive, show that it is described as if it were viewed by a regressive observer, and show that this imagined regressive individual would constitute a self-conscious human being. That will be conclusive evidence that we are self-conscious creatures who are using that regressive method of defining ourselves and our surroundings.

CHAPTER III

TABULAR ANALYSIS OF A REGRESS

The French philosopher Descartes, while engaged in subjecting all so-called knowledge to the acid test of doubt (in the hope of discovering something indubitable), was seized by a sudden inspiration. 'I am thinking!' he exclaimed, 'Therefore I exist.'

Critics have declared that this saying embodied two assertions concerning two empirical discoveries and that these findings should have been announced in the following order :

(1) 'There is thinking going on' (an undeniable fact, 'given' to introspective observation).

(2) 'This thinking is *my* thinking.'

For awareness of activities, and awareness that there is a 'self' which is active, are two very different matters.

Be that as it may, the initial fact which Descartes announced (before he brought in his unnecessary 'therefore') was: *I am* (thinking). And it is important to bear in mind that he was seeking, at the time, for something which *he* could regard as indubitable. So that he was regarding it as 'given' to him, without necessity of argument, that there was an 'I'—thinking. Thus, intentionally or unintentionally, he was claiming for 'self-consciousness' the status of given, undeniable knowledge.

We are, all of us, aware of our thoughts. We can watch, critically, the sequence of mental operations we are performing in any reasoned argument, so that an error is detected and arrested before the next step is made. We can retrace any train of ideas we may happen to have followed in mind-wandering. Indeed, it was only because a great part of our thinking processes —remembering and associating—are observable to introspection that the science of psychology came into existence.

But, if it is, for you—the present reader—an experimentally ascertainable fact that *you* can observe such thinking processes, this involves, not only your direct knowledge of the processes but also your direct knowledge of the something—called or miscalled 'yourself'—which thus observes them.

Now, if there be such a 'self', it is not readily discoverable by introspection. We seem to know of it, in fact, from the presented verdict of mental processes which we have been unable to follow. Yet the knowledge thereof is, certainly, 'given', in the sense that we cannot rid ourselves of it by any means whatsoever—not even by reflections on the obscurity of its origin.

Most people are prepared to accept self-consciousness as a fact; even though they regard it (wrongly) as a fact which plays no part in our interpretation of the physical world. But everyone finds it unsatisfactory to be confronted with something which claims the status of existence while declining to submit to examination. I suggest, therefore, that we make one more attempt to track down this elusive 'self'; and, since our powers of conscious introspection seem to be too feeble for this purpose, I propose that we set about our task in an entirely different fashion.

We shall begin by imagining that there exists a 'self-conscious' *observer*. He is to be aware of his 'self' as something observed. He is to distinguish that 'self' from an antithesis—a 'not-self'—also observed. And he is to be aware of his 'self' as an intermediary entity—an instrument—which he can employ in observing the 'not-self'. In other words, he is to be aware, by observation, of what is called 'the subject-object relation'.

Then we shall ask ourselves what sort of a thing such a creature would need to be in a rational world—a world which science could handle.

When we have ascertained those requirements, we shall look around to see whether there is, or is not, in nature as we know it to-day, anything which meets that bill.

We shall find that our bill of requirements constitutes an infinite series which we shall need to draft in the form of a table. The table will be triangular; consisting of an arrangement of compartments like this,

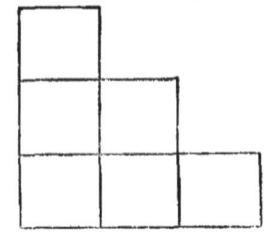

—which looks, at first sight, as if I proposed asking the reader to examine something much more complicated than the simple series of ancestors, or of whole numbers, we glanced at in Chapter 1. That, however, is not the case.

This tabular construction is only a convenient way of exhibiting the relations between all the 'terms' of any simple series. Let us glance at an example. We can realise, quite easily, that every schoolboy is the child of the child of the child of the child of - - - the remainder of an extremely long series of ancestors. But, if I were to ask you what was the relation between the second and fifth individuals in that series, you would have to think for a moment or two before you could reply that the one was the great-grandchild of the other. You would have to think much longer, if I asked you the relationship between the ninth and the thirty-second terms. But I could prepare for you a triangular table which would save you any trouble of that kind. And I should construct it as follows.

In the top compartment of the table I put the first person of the series, the schoolboy, as described by the second person, the father.

1st person
child

In the next (horizontal) pair of compartments I put the grandfather's descriptions of the first and second persons, the child and the father.

1st person	2nd person
child	
grand-child	child

In the next row I put the great-grandfather's descriptions of the child, the father and the grand-father.

1st person	2nd person	3rd person
child		
grand-child	child	
great-grand-child	grand-child	child

In the next row we include the great-grandfather, and give the great-great-grandfather's descriptions of all his descendants.

1st person	2nd person	3rd person	4th person
child			
grand-child	child		
great-grand-child	grand-child	child	
great-great-grand-child	great-grand-child	grand-child	child

And so on for as far as you like.

Please note that,

(1) Each row gives the relations which all the persons considered therein bear to the person on the extreme right of the line below. The last row gives, of course, the relations of the persons to the individual who comes next in the series.

(2) Since each row describes the persons concerned *as these would be described by the person next to be considered,* the descriptions change in each row. For example, the second person of the series (counted from left to right) is child in the opinion of the third person, grandchild in the opinion of the fourth person, great-grandchild in the opinion of the fifth person (not yet entered), and so on.

(3) The descriptions given of each person are only *characters* pertaining to them on account of their different relations to the different individuals of the series. We are trying, throughout this table, to arrive at a description of each individual as the descendant of the *ultimate ancestor.* When we arrive at the stage where we discover the great-great-grandfather, we declare that the person with whom we started is to be described, properly, as the great-great-grandchild of that ancestor. That definition is given in the left-hand compartment of the fourth row. This child's other descriptions (in ascending order up the first vertical column) are regarded then as merely *characters* which, we have discovered, are bound to pertain to any great-great-grandchild. Unfortunately, we cannot reach, in the space at our disposal, the ultimate ancestor; but we shall find that a great-great-grandchild, in turn, is only a character which must be possessed by a great-great-great-grandchild.

The reader need not trouble, here, to learn the ins and outs of this table by heart. He will have plenty of opportunity to familiarise himself with these as we go along. The essential thing now is for him to realise that the table is quite comprehensible, and that it deals with various aspects of only one simple series. Also, that the descriptions given are, all *relative—the* table does not tell us what anything is in itself. For instance, our first entry tells us nothing about the schoolboy—except the way in which he is related to his father; it describes him simply as 'child'. The other entries follow the same rule.

CHAPTER IV

REGRESS OF SELF-CONSCIOUSNESS

When we are trying to describe what we mean by self-consciousness, we say that you are aware of '*yourself*', that I am aware of 'myself', that she is aware of '*herself*', but that he is aware of 'himself'. This last is a bad error, for the possessive pronoun is all-important. There could be nothing rational in a Jones who was aware of Jones, and science could have no dealings with such an individual. You are speaking quite properly when you say that you are aware of 'yourself'—and not of 'youself'.

The only 'self' that you could be aware of, in a rational world, would be something which was an *object* to the ultimate, real you. But your self-consciousness does not lie merely in your being aware of such an object—it involves the recognition of that object *as yours*. Suppose you decide (rightly or wrongly) that your body is 'yourself'; you do not do so because you are aware of *a* body—a body belonging to, say, Smith—but because you are aware of the body in question *as yours*. And so it is with any subtler object you may designate by that title of 'self'. A man who was aware that 'he' was observing would be aware of an observing thing which was an object to the ultimate him; but, to be self-conscious, that man would have to be aware of that observing thing, not as an object apparent to the human race in general, but as an entity pertaining strictly to him. He would need to be aware of it as *his* observing self.

It is easy to see, now, that any rational self-consciousness would involve an infinite regress. For, whatever were observable to a man as a proper 'self' would need to be observable to him as *his* self, involving awareness of *something owning* the self first considered. Let us suppose, for example, that B is recognised by the self-conscious individual as his observing self and A as the object (the 'not-self') observed—an arrangement which we can tabulate thus,

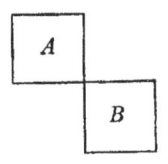

```
┌─────┐
│  A  │
│   ┌─┴───┐
└───┤     │
    │  B  │
    └─────┘
```

putting (for future convenience) the observing entity to the right of, and

25

below, the entity observed. Then, since the self-conscious creature regards *B* as *his* self, he must be aware of a self *C* which owns *it*. So that the table must be extended thus,

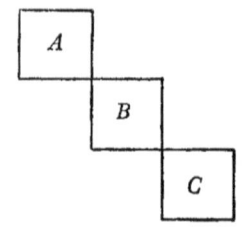

indicating that *C* observes *B* while *B* observes *A*. But, since our friend is aware of *C* as a 'self' owning *B*, he must be aware of that *C* as *his* self, and so be aware of a self *D* owning *C*, thus,

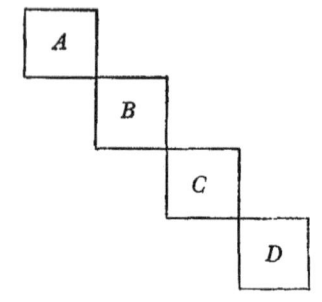

where *D* is observing *C's* observations of *B's* observations of *A*.

D, of course, must be a 'self' observed by an owner *E*, and so on *ad infinitum*.

It looks rather fantastical, as do all regressions when we first encounter them. But there is no getting away from it. Unless *D* is aware of *C*, he cannot regard *B* as *his* self—not, at least, in a rational world.

The reader, however, studying this table, will ask the following question : 'If *C* observes *B* while *B* observes *A*, how can *C* be aware of *A* as distinct from *B*? Surely he would observe *B's* response to *A* as merely a modification in *B'*. This criticism is quite justified. It is, indeed, the basis of the philosophy called Idealism—the theory which denies the separate existence of *A*.

We must recognise, then, that our table, though is incomplete. There is a great deal missing. And what that great deal is we shall discover in the next two chapters.

CHAPTER V

MEANING OF 'OBSERVATION' IN PHYSICS

Let A and B be two entities existing independently of each other. Let A be *affecting* (I am choosing the word with the broadest meaning) B. And let us suppose that we are studying the effect of A upon B. In making that investigation we are, actually, employing B as an instrument for discovering something about A.

Now, it is clear enough that the knowledge of A provided for us by B can be knowledge of only a single character possessed by A—the character of *being-able-to-affect-B*. This character is said to be 'relative' to B; since, by our definition thereof, it does not exist except with reference to B. But it cannot be the only character which A possesses; because, if that were the case, the complete A would be merely relative to B and have no independent existence such as we hypothecated at the outset.

Suppose we designate the fully charactered A by A_2, and represent the character of *being-able-to-affect-B* by A_1. Then what B, the instrument, is said to 'observe' is simply this A_1 — for characters of A_2 which do not affect B are, obviously, not discovered for us by B. The instrument B is referred to, in science, as 'the observer'.

Thus, in science, to 'observe' is to *abstract a character from* some entity existing independently of the observer. And the character abstracted must be one which, in some way, affects that observer.

We see, then, that an 'observing instrument' is not, in strict scientific parlance, a mere *measuring* appliance (though it may have a scale attached to it as a refinement). As examples of observation by an instrument, I may cite: A dynamometer abstracting Force from Impulse; a moving body with its motion restricted by the proximity of another body and which, thus, abstracts that other body's character of Attraction or Repulsion. All these abstractions could be made without the use of any scale to give a merely numerical magnitude to the character abstracted.

It is to be noted that, if our knowledge were confined solely to knowledge of B, we should have no grounds for supposing that B's behaviour was due to anything beyond its own intrinsic nature. Our science would consist then of a mere classified catologue of the incidents in B's career, and we should have no right to speak of B as an 'instrument'. The use of that term implies that we have some previous knowledge of A_2

as an entity other than the known B.

The knowledge involved in a scientific experiment may be classified, then, as follows.

Observed by (abstracted by) B	A_1	
Known to ourselves and regarded as existing independently of each other	A_2	B

It will be perceived that, from the outset, we credit B with a reality which we deny to A_1. For A_1's existence is merely relative to B. It will be realised, moreover, that it is impossible for us to regard an instrument B as something which we can add to a system consisting of entities (such as A_1) which have been described solely by the way in which they affect B.

CHAPTER VI

REGRESS OF A SELF-CONSCIOUS OBSERVER

We are now in a position to tackle the individual to whom it is a 'given' fact that 'he' is observing something which is not his observing 'self'.

Let A be the object observed, B the observing self', and C something which knows that B is observing A. These we can tabulate as before *(vide* Chapter III).

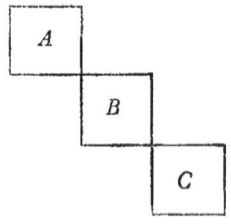

The question was: How can C be aware of A as anything but a modification in the B which he is observing?

We know from the last chapter that A, being something *observed* by B, is merely a character abstracted from some entity in the world which contains B. We can describe A, therefore, as an A_1 abstracted from an A_2, and can amplify our table in the fashion shown below. Since there may be any number of A_2 entities affecting B, we may call A_1, ' World as observed by B'.

World as observed by B	A_1		
	A_2	B	
			C

Now, since it is to be, for C, an unavoidable judgment that B *is* observing some character of A_2, he must have a knowledge of A_2 as much 'given' as is his knowledge of B, that is to say, it must be knowledge by observation. So we can fill in a little more of our table; thus:

World as observed by B	A_1		
World as observed by C	A_2	B	
			C

Now, since A_2 and B are observed by C, they must be characters abstracted by C from corresponding entities in some more fundamental world containing C the observer. So we can change B into B_1 and can tabulate the two more fundamental entities as A_3 and B_2; thus:

World as observed by B	A_1		
World as observed by C	A_2	B_1	
	A_3	B_2	C

Here, C is aware of an objective A_2, and of B_1 as an object which is being modified by the character A_1.

We know that, since B_2 is having its character B_1, modified by A_1, it is recording the presence of A_1. But to record the presence of A_1—the character of A_2—is not to record the presence of A_2 as a whole. A_2 as a whole, is *not* being observed by B_2, and B_2 is not abstracting A_2 from A_3 It is C who is doing that, i.e., A_2 is that character of A_3 which *is relative to C*, but it is not in any way relative to B_2.

But the regress of self-consciousness, which we studied in Chapter IV, declares that C itself is only a 'self' observed by a remoter owner, D, who is the real, ultimate observer of the series, as far as we have considered this.

Now, by our hypothesis, this (so-far) ultimate observer D has to know that A_2 *is* an object existing independently of his self B_1. Of course, C records, as we have seen, the separate existences of A_2 (containing A_1) and B_1. But these recordings are only modifications of, or changes in C. The question is, again, how can this ultimate observer D know that A_2 (containing A_1) and B_1 are existing independently of, and being observed by, C, and are not merely modifications in the structure of C.

D cannot discover that by merely observing C. The answer is that to discover that A_2 and B_1 are observed by C is to perceive that C abstracts them from some more fundamental entities. The entities from which C does abstract them are, as we have seen, A_3 and B_2. D, therefore, must perceive that A_2 and B_1 are abstracted from A_3 and B_2 by C. But, as a preliminary to observing this function of C, he must be able to observe A_3 and B_2.

So we can amplify our table by labelling the third row, 'World as observed by D'.

World as observed by B,	A_1		
World as observed by C	A_2	B_1	
World as observed by D	A_3	B_2	C
			D

Then, again, since A_3 and B_2 and C are observed by D, they must be characters abstracted from more fundamental entities, A_4, B_3 and C_2, in the same world as D. So we can change C into C_1 and extend our table thus:

World as observed by B_1	A_1			
World as observed by C_1	A_2	B_1		
World as observed by D	A_3	B_2	C_1	
	A_4	B_3	C_2	D

But the regress of self-consciousness insists that D, itself, is only a 'self' observed by a remoter owner E, and so on *ad infinitum.*

Clearly, then, if we wish to complete our analysis of an individual to whom it is 'given' that his 'self' is observing something, we shall have to extend our table to infinity, repeating the old arguments for each new entity introduced.

It is to be noted again that the abstractions are all performed by the series of observers B_1, C_1, D, etc., along the diagonal edge, and not by any other entities shown in the table. We saw, before, that B_2 does not abstract A_2 from A_3, and similar arguments will show that B_3 does not abstract A_3 from A_4, and that C_2 does not abstract B_2 from B_3. This rule must hold good throughout the infinite regress.

It is evident that, in the four-world table shown, there is only one world adjudged as being real—the world of the bottom row. The 'worlds' tabulated in the other and upper rows are merely lists of characters abstracted from that more real world by D employing the primary observing instrument C_1 and the secondary instrument B_1.

The character of the regress is clear enough. We have a horizontal series of entities, indicated by the alphabetical sequence A, B, C, etc., and a vertical series of characters of those entities, indicated by the numerals 1, 2, 3, etc. The regress of the self-conscious observer who is aware of an object A_1 other than his 'self' lies along the diagonal edge B_1, C_1, D, etc.

That the ultimate observer should be able to treat the series of entities A_1, B_1, C_1, etc., as independently existing systems is a condition essential to his possession of any knowledge of a 'self' situated in an external world. But that is only the half of our trouble. In order to fulfil our requirements, the observer in question must be able to recognise, not only that A_2 exists independently of B_1, but also that A_1 is being *observed* by B_1 ; which means that he must be able to perceive that the modification in B_1 *is caused by* the nature of A_2. And, similarly, throughout the regress, he must be able to perceive, not only the separate existences of the observing instruments and the systems from which those instruments are abstracting, but also the fact that the instruments are being affected by characters of those systems. Now, our present table does not show *how* the ultimate observer is enabled to perceive this : it merely assumes that he can do so. And that, of course, is insufficient for our purpose.

It will be realised that our test is very drastic. We have to discover, in our everyday, scientific methods of describing the universe, some *unnoticed assumption* which actually takes into account all that infinite series of different entities indicated in the horizontal extension of the table. In addition, this commonplace method of description has to make it clear that the ultimate observer will perceive the observing entities as observing and the observed entities as observed. And not till we have discovered this immensely significant assumption, and have shown that all our empirical sciences are founded upon it, shall we be in a position to assert that we are self-conscious individuals, aware of an external world, and employing the regressive method of the artist and the picture because it shows in a reliable and useful fashion the otherwise incomprehensible relation between ourselves and our universe.

That descriptive convenience exists. We put it to everyday use. And, if you like to say, in view of the enormous difficulty of the problem, that any such device would need to be the product of a master Mind, I, for one, shall not attempt to contradict you. But the greater marvel, I think, lies in the fact that the device which solves the tremendous problem of rendering systematic an otherwise incomprehensible world proves to be, at the same time, of such a character that the veriest half-wit, lacking all clear understanding of its nature, is compelled to employ it. The Mind which devises the method devises it for the advantage of both the genius and the fool.

THE SERIAL UNIVERSE

PART II

GENERAL TEST OF THE THEORY

CHAPTER VII

'NOW'

Let M represent a particular configuration of the external world as this last is described by you from observation, experiment and calculation. The particular configuration which M is to represent is the one which is open to your observation at the present moment. Let L represent, similarly, a past configuration remembered. From your knowledge of L and M you calculate, let us suppose, what will be the character of a future configuration N. Your descriptions are made in the language of classical science.

If, now, you examine your three descriptions, you will discover that these amount to no more than descriptions of three separate worlds. For there is nothing to show that one description refers to anything more or less real than does another. Equally, the descriptions give no indication that any of the configurations are past or present or future.

Further examination brings to light that the three worlds described differ from one another in the condition known to science as 'entropy', and that the nature of this difference is such as to allow you to consider these worlds as arrangeable in order of their amount of entropy (an arrangement which will correspond nicely with our alphabetical order LMN). *This* entropy order we may hope to describe, presently, (though we are not yet entitled to do this), as time order. So far, however, the descriptions fail to show,

(1) That they refer to *successive states* of one and the same world, or

(2) That those states have any relation to a 'now'.

As we shall see shortly, these two requirements are merely different ways of expressing the same thing. We cannot assume condition (1) without assuming condition (2). But we need not enter into that question here. It is sufficient, for the moment, to note that our descriptions do not fulfil condition (2).

Examining condition (2), we remember that M was to represent the configuration which is open to *your* observation 'now'. A doubt assails us here. For a great many people have supposed that the notions of a 'now' and of 'happening in succession' are references to a psychological observer which ought not to be made. The order exhibited in our present descriptions L, M and N, provides, it has been said, all that is needful for scientific purposes.

Very well, suppose we ignore the fact that the actual starting point of your description was your observational knowledge of M and your remembered knowledge of L. We have no shadow of right, of course, to do any such thing; but we are trying to put ourselves into the position of these objectors. Let us say that the reference to yourself as the observer—the reference which was implicit in the demand that M was to represent the configuration open to your observation at the present moment—was a reference which ought not to have been made. Let us say, if you like, that the 'now' is psychological—though classical psychology was as 'now'-less as classical physics. Let us say, even, (since we have lapsed into nonsense, and may as well be hung for a sheep as for a lamb), that the 'now' is an 'illusion'. Good. Our present description of L, M and N has been made by yourself from your memory, observation and calculation—we cannot avoid that—but it contains no reference to the observer and describer, and no unique 'now'. It is, in fact, the description which, according to these people, describes three temporal 'states', and which they assert to be entirely sufficient for the practical purposes of any man of science.

We must agree that it is very satisfactory to have arrived, by this drastic process of elimination, at a reliable account of the universe around us. But how can we be sure that it *is* reliable? Ah ! that is the beauty of science as distinguished from mere philosophy. We can test the truth of its assertions by actual experiment. Splendid. Let us test the accuracy of our present descriptions, L, M and N. Let us make an experiment and see.

The best configuration for us to employ for this purpose will be, I suggest, the one we have described as L; because, by experimenting upon (altering) that one, we shall be able to note whether configuration M is changed according to the calculated result, and to see, also, whether the change carries through to configuration N.

What's that you say? *We cannot alter L!* Why not? *Because L is past!* But we have just agreed that the world which we have described as L, M and N is devoid of such mystical characteristics as 'past' or 'present' or 'future', and that *this* is the world with which experimental science has to deal. What, then, is wrong with my proposal that we should experiment with the state L? *Something was omitted from that description!* Well, perhaps you are right. But what did we omit?

It needs no pointing out that any system which can be classified as an object to be experimented upon must be distinguishable—arbitrarily or otherwise—from the instruments which are regarded as interfering therewith for the purposes of the experiment and as measuring the results of that interference. The two systems must be treated as extraneous to each other. Now, the essence of a scientific description has been, always, that the validity of the description must be experimentally verifiable by everyone, including the describer. This limits the universe which can be described.

It must be one which the describer can regard as extraneous to his *instruments* and as subject to interference by these.

But, if the objective universe which is thus described is regarded by the describer as a series of 'states' possessing time order, it is, as we have just discovered, an essential condition that he regards his experimental apparatus (the excluded system which interferes) *as operative at only one 'state' in that apparent temporal series—the* 'state' he calls 'now'. And anyone who delegates to him the task of verification must agree with his verdict concerning which is that unique, assailable 'state'.

But how does the describer know which is this critical 'state'? What marks the 'now' for him? Is it physical as well as 'psychological'?

Consider this 'now-mark'. We know that it has a reference to the experimenter system. We know that it is a finger-post reading: 'This way to the interfering system which we left outside'. Consider, again, that we must regard this finger-post (whatever it may be) as *changing* from association with one configuration of the object series to association with the configuration which the describer regards as next in time order. Thus only can the mark indicate an important aspect of the problem, viz., that, if the experimenter system postpones its interference, it will find that its chance of altering the configuration which was 'now' has gone. The interfering-and-observing system follows, of course, these changes of the finger-post.

But, in these circumstances, the excluded instruments of the experimenter system, following the changes of the 'now', must *mark* that 'now' ! Quite so. And they constitute a *physical* 'now-mark' which the observer has made for himself. *For, when he extrappolates the observed system in time, he leaves his instruments, automatically at the psychological 'now'.*

When we have taken into account this behaviour of the 'now-mark' (the observer's instruments)—a behaviour indicating clearly that the series of configurations in entropy order, pertaining to the observed system, is being presented to the observer's instruments in *succession*—we shall be entitled to say that these configurations have been described, quite properly, as states successive in time—to those instruments.

And that is the truth about the time device as employed by all experimental science. It separates the observed and observing systems in the most effective fashion possible—by providing them with what are (as easily may be proved) *two different time systems interacting at a 'now'.*

Now, this simple fact about scientific analysis in terms of time—that a system which is accepted as obtaining information by experiment must be treated as an interactor which is (to use simple metaphors) 'travelling through' any 'time-map' which that acceptance allows us to draft—was not appreciated by the classical employers of the device. The fact itself is, evidently, a special example of the general law to which we directed our attention at the end of Chapter V, viz., that it is mathematically impossible to treat B (a thing which is affected) as an additional part of any system A_1 which is being described by the way it affects that B. The materialist, for example, would have argued that it is possible to add to the sequence of material states LMN three corresponding states of a system of material instruments, lmn, thus,

$$L \quad M \quad N$$
$$l \quad m \quad n$$

and to regard lmn as the system of instruments which provides the information from which the description of LMN as material is compiled. And the reply would be: (1) (On general grounds) That this would be to commit the mathematical fallacy of trying to put the observer into a temporal system which has been described by the temporal features it presents to that observer; and (2) That—as an empirical fact which is merely illustrative of (1)—the experimenting, interfering, pressure-exerting instruments which provide the information from which the material description is compiled must, of necessity, be treated by the describer as confined to a 'now': a state of affairs which he must represent thus,

$$L \quad\quad M \quad\quad N$$
$$O$$

where O is the instrumental system in question. If we ask : What, then, is represented by *lmn* in the materialist's picture? the answer is: The successive states of some piece of mechanism designed for use as an instrument but which *is not being employed by the describer as a source of information or as a means of interference*. In that picture *both LMN* and *lmn* are being described by the way they affect the describer's instruments, which last have not been shown.

In their actual work, all the men of science, guided by sound intuition, avoided the materialist's fallacy. They had no clear notion that they were relegating observer and observed to two different time systems, or that they were entertaining the idea of a material 'now-mark' changing from association with one state of the system observed to association with the next. *But they did this, unconsciously, whenever they separated the experimenter and his instruments from the system to be experimented upon, and accepted that experimenter's view of the object system as a series of states in time order. And they did that in every experiment they made.*

Before we go on, there is one rather remarkable fact to which we should direct attention. All this means that 'determinism' is 'non-suited'. Not only has it no case to present : *it never had a case.* Classical science involves, employs and asserts the contrary view—the view of every observer as an external potential interferer with an otherwise determinate universe. We need no microscopic 'Uncertainty Principle' to assist us there. The determinist bogey—that alleged offspring of classical science—was never even conceived, and the birth certificate signed by the materialist was a fake.

CHAPTER VIII

REGRESS OF TIME

We have, seen that time is an analytical device which effects the sharpest possible distinction between subject and object. We can see, also, that each person will apply it differently. Jones will regard the system upon which he is experimenting (which may include Smith) as a series of states in an objective time order, while he treats his instruments of observation and interference as confined to a 'now' which changes from association with one state of the system observed to association with the next. Smith will regard Jones (and Jones's instruments) as pertaining to the objective time series, while considering that it is his own instruments which are excluded and confined to a 'now'. Thus, Jones's instruments may be considered, in some cases, as belonging to a series of objective states, and, in other cases, as confined to a changing 'now'; according to whether we are employing Smith or Jones as our source of information. Obviously, then, analysis in terms of 'time' is merely a *mathematical convenience*. And it is one which gives the maximum prominence to the subject-object relation. We need not be surprised, therefore, if we discover, presently, that its mathematical character is regressive.

In the last chapter, we represented the three distinctive entropy configurations by three letters, L, M and N. This was in anticipation of the later stage where we should be able to regard those configurations as successive in time—to the observer's instruments. The alphabetical *sequence* of the letters would serve, at that stage, to indicate the order of *succession of* the states of the observed system. Now, although we may, for convenience, write the letters in a row, it must be understood that this positional arrangement is not essential to the argument. We could, if you preferred it, write the letters on counters and shake these up in a bag. The entropy order which indicated the time order would still be indicated, quite adequately, by the alphabetical order of the letters.

We have seen that the 'now-mark' which indicates the presence of our experimental instruments must be thought of as changing from association with one state of the system observed to association with the next in whatever represents, to those instruments, the order of objective events. In the state of affairs we have been imagining as confronting us, we assumed the 'now-mark' to be at M, thus,

$$L \qquad \text{\textcircled{M}} \qquad N$$

the mark being represented here by a circle enclosing the significant letter. In this state of affairs, M is present, L is past, and N is future—to the instruments in question. But we may not think of the mark as remaining indefinitely at M, allowing us as much time as we desire to prepare for an experiment on the basis of that present state of affairs. A little later on we shall find that the mark is associated no longer with M. We may have to represent that future state of affairs thus,

$$L \qquad M \qquad \text{\textcircled{N}}$$

where N is present and L and M are both past to the instruments concerned.

Again, we have realised that we cannot experiment with L, because L is past (to the instruments). But we have to recognise that there was a past state of affairs where L was present and M and N were both future (to the instruments), a state which we may represent thus:

$$\text{\textcircled{L}} \qquad M \qquad N$$

Now, what precisely did we mean when we said that $\text{\textcircled{$L$}} MN$ represents a 'past' state of affairs, that $L\text{\textcircled{$M$}}N$ represents the 'present' state of affairs, and that $LM\text{\textcircled{$N$}}$ represents a 'future' state of affairs?

Let us label these three states of affairs *1, 2* and *3,* and let us place them (for convenience) one above the other, thus:

```
3.      L           M          Ⓝ
2.      L           Ⓜ           N
1.      Ⓛ           M           N
```

We know that M represents that entropy configuration of the observed system which we regarded originally as 'present', and that we accepted L as 'past' and N as 'future'. For that reason we placed the 'now-mark' at M. But we are realising now that this mark has changed from association with L to its present association with M, and is going to change to association with N. We intend, therefore, that Ⓜ shall indicate the present state *of the 'now-mark'* (i.e., of the instruments). Similarly, we intend that Ⓛ, indicate a past state, and & Ⓝ a future state *of the 'now-mark'*. But these intended past, present and future states *of the now-mark,* Ⓛ, Ⓜ and Ⓝ are being regarded as successive in a time order which can be represented only by our numerals *1, 2* and *3*!

Clearly, in *that* time order, the three states of affairs *1.*Ⓛ *MN, 2.L*Ⓜ*N* and *3.LM*Ⓝ represent successive states of a more comprehensive system.—a system which includes the three object states *L, M* and *N plus* the changing 'now-mark'.

Now, if M is to be present to the instruments, Ⓜ must represent (as we have just said) the present state of the 'now-mark', and this means, in turn, that *2.L*Ⓜ*N* must represent that state of the more comprehensive system which is present in the more fundamental time order indicated by the numerical sequence *1, 2* and *3*. But our descriptions do not indicate this ! For all they tell us, *1.*Ⓛ *MN* or *3.LM*Ⓝ might indicate, equally well, the present state of this circular mark. Clearly, then, we must add to the states *1, 2* and *3* of our more comprehensive system a new 'now-mark' indicating that *2.L*Ⓜ*N* is present in the more fundamental time concerned. We can do this by enclosing *2. L*Ⓜ*N* within an oblong, thus:

```
    3.      L           M          Ⓝ
  ┌─────────────────────────────────────┐
  │ 2.      L           Ⓜ           N  │
  └─────────────────────────────────────┘
    1.      Ⓛ           M           N
```

It is clear enough that the time order indicated by *1, 2* and *3* is more fundamental than the merely apparent time order which we indicated by the alphabetical sequence *L, M* and *N*. For it is the association of the oblong 'now-mark' with *2.L*Ⓜ*N* which makes Ⓜ the present state of the circular mark *0,* and which, thereby, indicates *M* as the `present' state of the originally considered system. If the oblong 'now-mark' were to enclose *3.LM*Ⓝ then

(N) would be the present state of the 'now-mark', and N the 'Present' state of the originally considered system—despite the fact that M in 2 is enclosed also by a circular mark.

It will be asked : Since we are trying to regard real time order as represented by the succession of the more comprehensive states *1, 2* and *3*, what was indicated by the entropy order of the original, less comprehensive configurations L, M and N?

Are we to try to imagine the more comprehensive system as embracing two kinds of time?

Certainly not : the more comprehensive system possesses only one time order, viz., that indicated by *1, 2* and *3*. It contains, also, all present, the items of the order indicated by L, M and N; but that order is not time order—in the more comprehensive state of affairs. Then what sort of order is it? Well, I am going to answer that question in the next chapter; but I have a particular reason for not wishing to do so here. In this chapter I am concerned to show *only* that real time order is the receding element in an infinite regress. As such, we shall be coming continually upon orders which serve the purpose of time order for the particular stage we happen to have reached in the regress, but which turn out to be something different from time order when we get on to the next stage, just as each 'child' in the fictitious ancestry regress turns out to be 'grandchild' in the next stage. But what that 'something different' is in the case of time, we need not consider at this moment.

Before we go any further, we had better note that the placing of our three second-term states of affairs *1, 2* and *3* one above the other on the page is in no way essential to the arguments we have used. These would proceed in precisely the same *way if* L, M and N had been written on counters shaken up in a bag. We should have required three such bags to represent the three distinctive states of affairs where the circular 'now-mark' surrounds, respectively, L and M and N. And, to make the bag containing (M) represent the present state of affairs, we should have had to label the three bags; *1, 2* and *3*, and then add another label, representing a second-term 'now-mark', to the bag marked *2* and containing (M).

To prove that real time order recedes in an infinite regress, we have to show that the arguments which led us from first-term time to second-term time are bound to repeat themselves thereafter.

We have arrived at a system containing three second-term states, of which states, number 2 is surrounded by an oblong 'now-mark'. We represented the total system thus,

3.	L		M	(N)
2.	L		(M)	N
1.	(L)		M	N

and we noted that it is the presence of the second-term, oblong, 'now-mark' which makes (*M*) in 2 (instead of (*L*) in 1 or (*N*) in *3*) the first-term 'present' configuration with which we started. State 1 is, thus, past, and state *3* is future. But we have agreed that (*N*) will become, in a little while, the first-term configuration which is thus uniquely 'now'. But, for (*N*) in *3* to become thus uniquely 'now', the second-term, oblong 'now-mark' must change from association with *2* to association with *3*. States *1* and *2* will be then both past. Again, (*L*) was once 'now', and the oblong 'now-mark' must then have enclosed *1*. (*L*) *MN*. States *2* and *3* were then both future. Consequently, we are confronted with three different states of the *whole collection* of letters and 'now-marks' so far dealt with—snnthree states each containing *1, 2* and *3 plus* an oblong 'now-mark', but with this mark associated respectively with *1, 2* and *3*. And one of those third-term states (the one where the oblong 'now-mark' encloses *2*) will have to be enclosed in a third-term 'now-mark'. Or (to employ the other method) we shall need three sacks, each containing three bags with counters, with a 'now-label' on one sack, a 'now-label' on one bag in each sack, and a 'now-mark' on one counter in each bag, in order to show that one unique counter of all the lot represents the first-term 'present' state with which we started. And so it must go on—*ad infinitum*.

It is to be noted, particularly, that nowhere in the analysis of this regress have we introduced a new hypothesis. We do not state that the first-term series *LMN may* be the present state of a more comprehensive system: we show that it *must* be so. We show, in brief, that the entire regress was *implicit* in our opening statement that *M* is the 'present' state of three states of the observed system. That, of course, is a characteristic of all regressions: they do not proceed by *adding* new terms, but by showing that the existence of one term with a dual character involves the existence of all the remainder.

We are going to abandon, in a little while, the method of representing our series of states by letters of the alphabet or by numbers. We shall represent the original states by dots, and their intended time order by the space order in which those dots are placed in the page. That, of course, is the conventional scientific way of picturing time. We shall represent the changes of the 'now-mark' by changes in its *position on the page* ; that is to say, we shall imagine it as *moving* over the row of dots representing first-term temporal states. This is a far easier method of studying our present problems. But it begins by what a few people would regard as begging a question. Is it legitimate to use space order for our first attempt at representing time order? Actually, the answer is, yes; but the point is a very subtle one, and many people who have not gone deep enough into the matter would answer, no. Such persons might then proceed to the further error of supposing that the entire regress arose from our having begun by trying to represent time in an erroneous fashion. It is to avoid that objection that the present chapter has been written. The Bergsonians (the people with whom we are arguing) admit that states of time are distinctive and successive, but deny that they can be regarded as separated in the way that points of space are separated. Very well, our original descriptions of the three distinctive configurations L, M and N do not indicate that these are *separated*. For all that the descriptions tell us, we might be dealing with three configurations pertaining to three different worlds imagined by three different people. We intended, of course, that our descriptions should convey more than this; but we found that they failed to do so. They exhibited merely an adaptability to classification in terms of entropy order.

Next, we note that nowhere have we used space order to represent time order. It is true that the counters in our bags are spatially separated, but their space orders in the bags may be changed as often as we please (by shaking the bags) without this affecting the alphabetical sequence corresponding to that entropy order which we hope to be able to regard as sequential—successive in time.

Next, we were particularly careful not to say that the 'now-marks' *moved* from one state to another; for to do this would have been to declare that the states were being *presumed* to be spatially separated. We said, instead, that the marks 'changed from association with' the next in whatever series we had been hoping, previously, to regard as real time order. That change, again, was not presupposed; it was discovered to be an empirical fact that our chance of interfering with any particular 'state' of the object system would vanish, and would be replaced by a chance of interfering with the 'state' which came next in what we were trying to regard as time order. It may be urged that the admission of this behaviour of the ' now-mark' is an admission that the states are separated in the same way that points in space are separated. Quite so. But this new view of the relationship between the states is a *development* of our original, less explicit view—a development forced upon us by the logical development of the regress. The new view is one which we have endeavoured to avoid, and had successfully avoided up till that moment. It is a *consequence* of the regress, and not a primary supposition *causing* the regress.

Finally, suppose we think of the distinctive state L as changing into the distinctive state M, while thinking of the observing entity outside the system as changeless (except when observing). Is that the same thing as thinking of the observer as changing from association with the state L to association with the state M? Of course it is. We are thinking of the states L and M as being successively associated with the unchanging observer; and it comes to precisely the same thing whether we say that L and M are successively associated with the observer, or that the observer is successively associated with L and M.

CHAPTER IX

REGRESS OF 'REALITY'. REGRESS OF PHYSICS. SPATIAL REPRESENTATION OF TIME

You will remember that we began by saying that M was to be our description of the state which is open to your observation at the present moment, and that L and N were to be described from memory and calculation respectively. According to popular notions, those descriptions should have shown M as real and L and N as unreal. They did not do so. They exhibited only three differing conditions of entropy with no reality distinction between them. Equally, the descriptions gave no indication that any of those conditions were present or past or future. Putting in the 'now-mark' at M rectified the latter deficiency. But did this addition reduce L and N to descriptions of the unreal states contemplated in the popular view?

We had best, I think, call upon one of the *exponents* of this common opinion and ask him what, precisely, is he trying to assert. His answer is as follows. M is a state which exists 'now'. L is a state which did exist once but does not exist 'now'. N is a state which will exist but does not exist 'now'. To say that states do not exist 'now' is to say that they are 'now' unreal.

We reply to this by asking him to which 'now' is he referring. Of course L and N do not exist at the first-term 'now'—we have been at pains to show that. But, certainly, they exist all right in the second-term 'now'.

This does not satisfy him. He suspects that our arrival at the second-term 'now' depended in some way on a presumption that L and N were existing states (which, of course, would have been begging the question). If, he thinks, we had been quite clear about the non-existence of these states when we referred them to the first and only 'now' we recognised at that stage, the regress would not have developed.

He is quite wrong. Let us suppose that the first-term 'now-mark' is, as he wishes, a mark conferring reality on the state described. Good: M describes a state which is real; L and N are descriptions of unreal states—unreal simply because they are not existing now. But, by his own account, N will be—in a little interval of absolute time—the description of a real state existing 'now', and L was once the description of such a real state. Analysing this conception, we find that it is simply the concept of our second-term state of affairs,

3.	L	M	(N)
2.	L	(M)	N
1.	(L)	M	N

where second-term time is real time, and first-term time is only a pseudo-time. Here, *1* contains descriptions of three states which are all past, while *3* contains descriptions of three states which are all future—in absolute time. Consequently, none of those six states is real.

But *1* and *3* each contain the original 'now-mark' which was regarded by our friend as conferring reality. So this mark has lost its supposed potency. It does not represent anything beyond *a description of the observer's three-dimensional instruments* —and it gives three descriptions of these; viz., as past or unreal, as present or real and as future or unreal—in real time. But the recognition of them as present or real (in *2*) is not due to anything distinctive in their description: it is due solely to the fact that *everything* in *2* is defined as a description of some state present in real time.

So this popular definition of reality regresses. And that means that it is only a definition of *relative* reality. It means that the state *M* seems real to the instrument—simply because it is the state which is being observed by that instrument. But that *we* regard it as real depends, obviously, upon whether we are regarding the instrument as real. And the nature of the regress is such that, *when we are regarding the instrument as real, we are regarding as equally real all states which are past or future in first-term time.*

It may seem strange that an attempt to regard the past and future as unreal should break down in this hopeless fashion. But the fact is that nobody actually has ever thought of them as unreal. We think of them merely as 'having been real', and fail to notice that this is thinking of them as real in what we are regarding as real time.

We have arrived at a satisfactory account of the man-in-the-street's views; but we must attend now to an interruption by a physicist of the old school. 'You admit', he says, 'that this first-term reality of yours is relative to the instrument. Well, that is the only kind of reality in which I am interested. I do not even consider whether my instruments are real—they are outside my picture. That picture is concerned only with what it is that affects the instruments.'

We will allow him to maintain this view, but only on one condition. He must agree that this 'real' world which he is examining with his instruments is one which he has never tested by experiment—has never altered. If he has altered it, it is a world in which his instruments have played a part other than that of mere observation; and to account for the present state of that world is to take into consideration the extent of the interference by the instruments—to consider, that is to say, the quantity of energy they have supplied. I think our classical physicist will prefer to bide his time and look for some weaker point of attack. Meanwhile, we may ask him to consider whether, if he contemplates any further experiments, he is regarding the future of his world as *stable* or *unstable*. (I prefer those words to 'certain' and 'uncertain', which do not mean precisely the same thing.)

But here is a modern physicist with a perfectly legitimate question. 'Do you', he asks, 'regard this second-term "now-mark" of yours as physical or merely as psychological? If the latter, it has nothing to do with my science and I am not compelled to take it into account. I can see, of course, that if I have to recognise it, I am launched, past all saving, upon an infinite regress. But you must not expect me to take this critical step except under dire compulsion.'

I am afraid that compulsion is there. Glance back at the last diagram. The circles enclosing L in *1*, M in *2* and N in *3* represent past, present and future states of the interfering instrument. To make your experiment, you must, at some time or other, do something to your instrument; you must move, at least, some of its parts. But you cannot alter a past state of the instrument: you can act upon it only when it is in the state which you regard as present. Consider what that means. You can alter the instrument in *2.*$L\,\widehat{M}\,N$, associated with the second-term 'now-mark'; but you cannot alter it in *3.* $LM\,\widehat{N}$ until (in absolute time) the second-term 'now-mark' has changed to association with this state of affairs. Thus, the second-term 'now-mark' represents to you *a facility for moving the instrument*. The increase in the instrument's momentum results, in the course of the experiment, in an increased momentum of the original object system. So, the second-term 'now-mark' is a facility for adding momentum to the original object. Such a facility must be physical; and the physicist is obliged to take it into account for the same reason that compels him to take into consideration the instrument—viz., because it is a cause of the observed behaviour of the external world.

Before, however, we attempt to elucidate in greater detail the physical aspects of the time regress, it would be advisable for us to see whether our present analytical method is not open to simplification. Our treatment of states and 'now-marks' has been, so far, entirely algebraical—a matter of the manipulation of five signs, viz., L, M, N, O and _____ . The spatial order in which we have distributed these signs upon the pages of the book has had no significance of any kind. But most algebra is amenable to simple pictorial illustration, and we may as well make use of this fact in the present case. Readers who do not like diagrams may, however, continue to employ our past method of treating these problems: our diagrams will introduce nothing that cannot be expressed by continued combinations of algebraical signs. But those combinations would become immensely complicated.

Our three original states of entropy L , M and N exhibit what is called 'betweenness' order. M comes 'between' L and N; and this holds good even though M be merely a more broken-up L , and N be merely L in a greater condition of internal mixture. Now, we can think of intermediate conditions between L and M and between M and N, and of intermediate conditions again between the five states thus considered. And we can continue this process indefinitely. We do not need, however, to carry it so far as to produce an *infinite* number of states. Before we reach such a condition of affairs we shall have arrived at a curious mathematical phase in our process; we shall have unearthed the notion of what mathematicians call a 'limit'. Then we shall be able to regard our immense number of states as constituting what we can recognise as a *'Continuum'.*

Consider, now, the first and second terms of our series of more and more comprehensive systems. We can tabulate these as follows:

Present in the system apparent to the observer's instruments	A_1 M	
Present in the more comprehensive system known to us	A_2 LMN	B The observer's instruments

In the second-term system, *L*, *M* and *N* are being treated as:
(1) Of the same class (entropy configuration).
(2) Equally real.
(3) Parts of a *continuum*.
(4) Equally present.
(5) Associated with an independently existing thing which changes from one to the other. And, if you wished to describe three configurations as separated in *space, you* could say no more about them than we have said of *L*, *M* and *N*. *

So, although we begin by using the entropy order of the three configurations to represent their *time* order, the result is the discovery that, in the second term of an inevitable regress, this entropy order represents order in an unsuspected dimension of *space*. And it is clear that *whatever* we may select, at the outset, to represent time order must represent, at the next stage, nothing but space order. In other words, time order must change to space order at each stage of the infinite regress of real time.

I shall raise no objection if you prefer to speak of this new dimension as 'configuration space', meaning thereby a mathematical device to be distinguished from the 'ordinary' three-dimensional space of the first-term system *M*. It is part of our argument that analysis in terms of time is a purely mathematical device. The essential thing is to recognise that this space, 'configuration' or otherwise, is *space* and not *time* in the second term of the regress. And, as such, the 'betweenness' order of *LMN* therein may be adequately represented by the positions of three dots on a sheet of paper, while the 'now-mark' may be represented by a fourth dot, superposed on M, and with its presence indicated by a letter *O*, thus:

$$\begin{array}{ccc} L & M & N \\ \bullet & \bullet & \bullet \\ & O & \end{array}$$

This represents the 'betweenness' order of *LMN*; but that is not enough. We have to indicate also that the arrangement will appear to the observer's instruments *O* as time order. In other words, we must show that *O* will regard *LMN* as a *sequence* in which *L* comes first and *N* last. That condition, however, is satisfied if we add an arrow to show the direction in which *O* is *moving* along the newly-discovered dimension, thus:

```
    L         M         N
    •         •         •
              0
 ——>
```

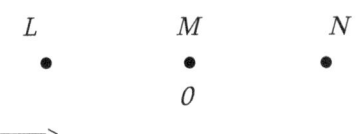

We have now an excellent graphical representation of first-term time order. But we have not yet shown that the three configurations *L*, *M* and *N* represent three successive states of *one and the same world* external to *0*. We have to introduce the notion of *continuity*. This we can effect by drawing a continuous line from *L* to *N*, thus:

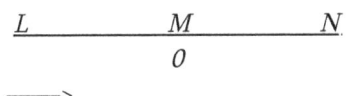

```
    L_____M_____N
              0
 ——>
```

Then any point in that line will represent one particular configuration of the world external to 0, and the whole line will represent the *endurance of* that world in first-term time.

Since the time sequence *of* these states is indicated by the arrow, we can abandon the alphabetical sequence *of* the letters *LMN* as redundant. A line labelled, say, *GH,* with an 0 somewhere between *G* and *H* to indicate the position of the ' now-mark', and an arrow to show the direction of its travel, thus,

```
    G————————————H
              0
 ——>
```

will be ample for our purpose.

And that is the method which was adopted when the time regress first was analysed. This was effected in a book called *An Experiment with Time,* published in March 1927. The method has great advantages of simplicity, and we shall employ it for the remainder of our present demonstration.

CHAPTER X

DIMENSIONS, MAGNITUDES AND MESH-SYSTEMS

I must ask permission to make a digression. The present reader, no doubt, is well acquainted with the meaning of the word 'dimension'. But I have in mind a potential peruser of these pages who happens to be a little hazy in his ideas on this subject. The digression is intended for his benefit.

A dimension is neither a line nor—strictly speaking—a magnitude. It is a *manner* in which something may be measured. For example, *'momentum'* consists of 'mass' multiplied by the velocity with which that mass moves. Consequently, it has to be measured in two totally distinct ways—one dealing with the mass and the other with the velocity. It possesses, therefore, two dimensions. We could say that mass and velocity are the two dimensions of a momentum, even though we did not know the amount of mass or the amount of velocity possessed by the particular momentum we are considering. Those amounts would be the *magnitudes,* and would need to be indicated by numerical figures; whereas the two *dimensions* can be represented simply by the symbols M (meaning mass) and V (meaning velocity).

Spatial dimensions provide us with a very convenient way of representing other dimensions. For example, we can employ the up-and-down dimension of this page to represent mass, and the side-to-side dimension to represent velocity. To indicate the amount of mass, we need a line OY laid down somewhere in the up-and-down dimension and marked off with a scale representing units of mass. Similarly, to indicate the amount of velocity, we need a line OX laid somewhere in the horizontal dimension and marked with a scale indicating units of velocity. But to employ the two dimensions of the paper to indicate the amount of momentum, we must place the two scales so that they meet at a common point O, and start the scale measurements from that point; thus:

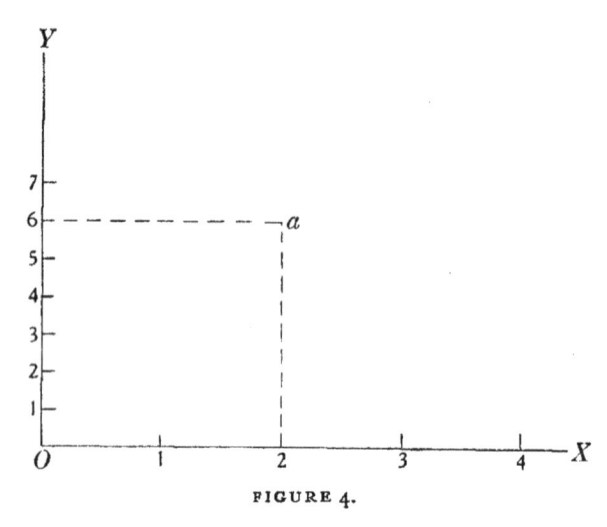

FIGURE 4.

The two scales OX and OY are called 'axes', and the point O at which they meet is called the 'origin'. You will notice that I have made the divisions on one scale quite different from those on the other. It is often a matter of pure convenience what sized scale you choose to employ in each case.

Now, consider any point a, placed in the angle between the two lines. The height of that point above OX, that is to say, its distance from OX in the up-and-down dimension, will give you a measurement of mass. You discover the amount of this by drawing a line through a, parallel to OX, to cut the scale on OY. In the present case, the mass magnitude thus indicated is 6. Again, the horizontal distance of a from OY will give you a measurement of velocity, the value of which you ascertain by dropping a perpendicular line from a to cut the scale on OX. The velocity magnitude indicated in this case is 2. Thus, the position of *a* with regard to the two axes indicates a mass magnitude of 6 and a velocity magnitude of 2, that is to say, a momentum magnitude of 6 x 2 = 12. The two magnitudes of mass and velocity (viz., 6 and 2) are called the 'coordinates' of the momentum.

The trouble about this dodge of using the dimension of surfaces to represent dimensions of other kinds is that the surface has only two dimensions available for the purpose. We can use a drawing in perspective to indicate a third—you can imagine, that is to say, a third axis sticking out from the page towards your eye—but this is a rather cumbersome device; and, when the dimensions with which we have to deal exceed two in number, it is more convenient to choose the two of these which you wish most to represent diagrammatically, and to refer to the others by letters of the alphabet. The treatment of those others is, of course, algebraical, while the treatment of the chosen two is pictorial; but this combination of treatments is quite easy and quite legitimate, since the diagrams are, really, only pictorial algebra. The point to be grasped here, however, is that, if you have to deal with something possessing a hundred dimensions, you can select *any* pair of these for pictorial treatment—the twenty-first and the seventy-fifth, for example, if it suits you—sticking to algebraical treatment for the remainder.

Let us return now to FIGURE 4, and let us draw through each of the divisions of the scale on *OY* a line parallel to *OX*. If we draw then through each of the divisions on *OX* a line parallel to *OY*, we shall have a network of crossing lines, as below.

This arrangement is called a 'mesh-system'.

You will notice that the two lines we drew from *a* in FIGURE 4 (a horizontal line to 6 in *OY*, and a vertical line to 2 in *OX*) were really two of the crossing lines of the mesh-system shown in FIGURE 5. In fact, FIGURE 4 was Simply FIGURE 5 with a lot of the lines of the mesh-system left out—for purposes of clarity.

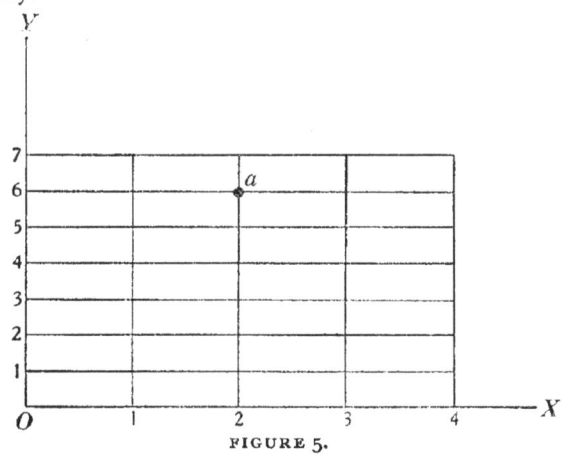

FIGURE 5.

Now, suppose that our axes of *X* and Y meeting at *0* were used to indicate, not measurements of mass and velocity, but measurements of distance in space. Distance from what? Well. look at

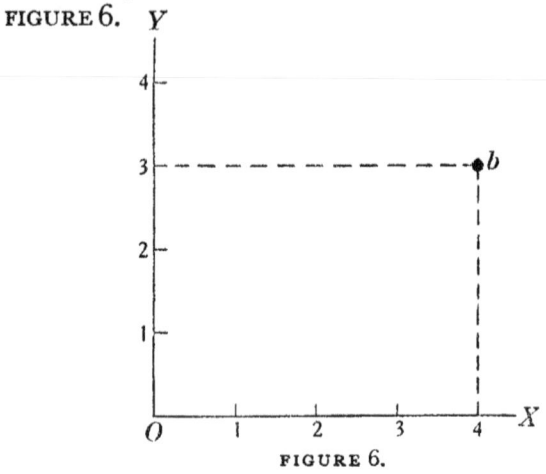

FIGURE 6.

FIGURE 6.

Clearly, the scale on OY shows that the point b is 3 space-units distant from the axis OX, while the scale on OX shows that the point in question is 4 space-units from the axis OY. Conversely, if we were told that the coordinates of some other point c were 3 in the horizontal dimension and 2 in the vertical dimension, we could place that point on the paper by drawing a vertical line upward from 3 on the X scale, and intersecting this by a horizontal line drawn from 2 on the Y scale; thus:

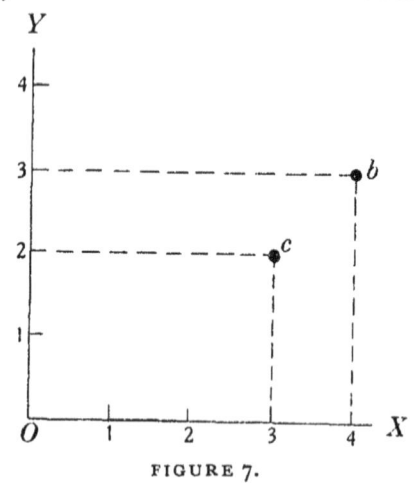

FIGURE 7.

With the aid of the readings on the two scales, and a little knowledge of elementary Euclid, we can calculate the direct distance in space-units between the two points c and b. But, if we propose to make the calculation, we must make the divisions representing inches on the *Y* scale equal to those representing inches on the *X scale.* Consequently, the meshes of the mesh-system—supposing that we fill this in—will consist of four-sided figures with all the sides equal.

Let us consider, next, a diagram which is fairly common in this era of influenza, viz., a temperature chart. Here we are using the scale on the vertical axis to indicate the height of the mercury in the thermometer (a space measure), so we may call this axis, the axis of *S (S* standing for space). The scale on the horizontal axis indicates periods of time as told by some clock, and we may label this axis, *T.* Here is one such chart.

It seems to indicate malaria rather than 'flu, but that is immaterial to you and me. The point I want you to notice is that I have made the vertical

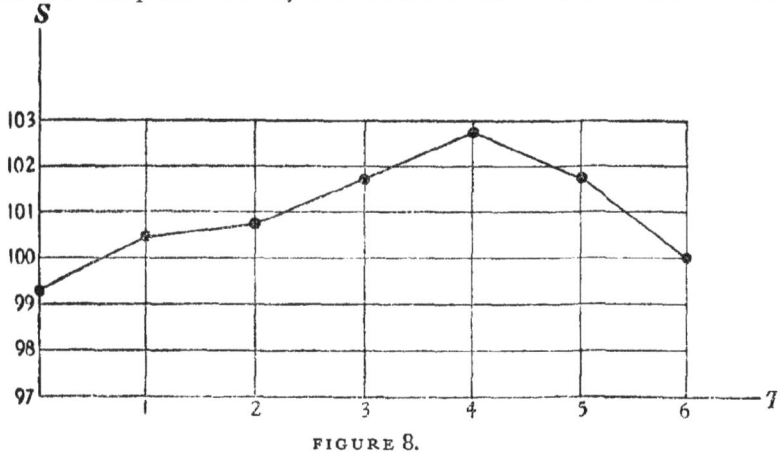

FIGURE 8.

spaces in the mesh-system much smaller than the horizontal spaces, and that *this is* immaterial. That is because the doctor is not interested in determining the *lengths of the lines joining the points,* but wishes to know only what was the height of the thermometer at certain instants of time. Any mesh-system will serve to inform him of this.

The nurse shows by round blobs the points where the patient's temperature was actually taken, and the lines joining the blobs are largely matters of guesswork. It is precisely such a line, however, which is called, in relativity parlance, a 'world-line'. Now, the relativist is particularly interested in determining the lengths of portions of such a world-line by methods which bear some analogy to the Euclidean calculation referred to earlier. Consequently, the nurse's mesh-system is not the sort of thing he likes. He prefers to make the space divisions of *his* mesh-system equal to the time divisions. How he contrives to make a period of time equal to a length of space is a matter we may discuss later.

Before proceeding with our analysis, it will be advisable to remind ourselves of a fact which was recognised by physics and philosophy long before Einstein embodied it in his greater 'relativity'—the fact that all measurements of velocity are relative to something. Now, the observer's instrument for determining the velocity of anything in the system observed can record only such velocities as are relative to *that instrument.* Suppose, then, that the observer, employing such an instrument as his source of information, prepares a space and time diagram exhibiting the spatial positions of the various parts of the observed system at different instants of time. The world-lines thus constructed will show, of course, by their inclinations to the axis which indicates time, whether the objects to which they refer are being regarded as moving in space or as at rest. And the time axis will represent the *track,* along the time dimension, of the observing instrument. The instrument itself is not shown —the diagram is a space and time map of the entities of the observed system only.

CHAPTER XI

GRAPHICAL ANALYSIS OF THE TIME REGRESS

Let us return now to our graphical representation of that second-term, more comprehensive system which includes the successive states of the first-term system plus the observer's instruments. We had discovered that the series of states of the original observed system could be treated as cross-sections of a continuous line *GH* representing the endurance, in first-term time, of that system, as in the figure below,

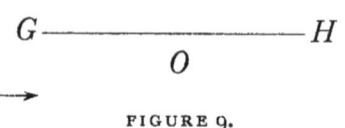

G————————————H
　　　　　　O
　→

FIGURE 9.
(FIGURE 8 of *An Experiment with Time,* first edition.)

and that the observer's instruments—the physical 'now-mark'—could be represented by a point *O* superposed on that line somewhere between G and *H*. We had ascertained that the actual world which we represent by *GH* must be thought of as extended in a hitherto unconsidered dimension of *'space** (a fourth dimension), that the observer's instruments represented by *O* must be regarded as travelling in that dimension, (the direction of travel being indicated* by the arrow in the diagram), and that the points on *GH* must be considered, consequently, to be appearing to the observer's instruments as the successive states of an ordinary, three-dimensional world.

It will be remembered that these instruments are interfering instruments exerting force upon the object system and, so, are observing, by reaction, that inertia which is the characteristic of *mass.*

The states represented by the points in *GH* are supposed to be described by

---

* We shall see, later on, that this involves no contradiction of relativity.

us from information obtained by use of the instruments at *O*. The process is somewhat analogous to that by which a man, having thrown, through a narrow vertical slit, a searchlight beam upon a dark external world, has prepared, from the glimpses thus obtained, a map of a countryside through which, he judges on other grounds, the searchlight and the slit, contained in a railway carriage, are passing. The analogy assumes that the man can estimate, from what has been seen, the probable character of the country to which he is coming; but, that much being allowed, it is obvious that he could both prepare his map and mark upon it the present position of the travelling searchlight.

The successive states of our second-term world will consist of a series of pictures like FIGURE 9, with the 'now-mark' *O* at different places in each picture. FIGURE 9—the whole of it—will be the present state of this more comprehensive world. States where *O* is nearer to *G* will be past states and states where *O* is nearer to *H* will be future states—in what we are regarding now as real time. Here the arguments of Chapter IX repeat themselves. The states of second-term time, showing the successive positions of *O* as this travels along *GH*, possess 'betweenness' order; and may be exhibited as a *continuum* (which is, of course, only a way of showing that the motion of *O* along *GH* is being regarded as continuous). Then, since the second-term 'now-mark' represents, as we saw earlier, something to which we are obliged to give a physical significance, and since this physical thing is changing from association with one part of the new *continuum* to the next part in order of continuity, we may represent second-term time by a dimension of space over which the second-term 'now-mark' is travelling. We have to consider, however, that three dimensions of space are reserved for 'ordinary' space in which the parts of the object system have different positions at different instants of first-term time, and we are considering that a fourth dimension of space is being employed to represent first-term time order. Consequently, the new *continuum* in which we indicate second-term time order will necessitate our employing *a fifth* dimension of space. The surface of our paper will allow us to represent this very nicely; the side-to-side dimension representing, as before, the fourth dimension, the up-and-down dimension representing the fifth dimension, while the three dimensions of 'ordinary' space are left out of it for algebraical treatment.

Here, however, a difficulty confronts the printer of the book. Strictly speaking, we should begin with the representation of our second-term world by the line *GH* and the point *O* as in FIGURE 9. That would indicate the present position of the first-term 'now-mark' *O*. We should then draw similar horizontal lines below this line to represent past states of this second-term world, (with *O* nearer to *G*), and we should draw another set of lines above *GH* to represent future states (with *O* nearer to *H*). But, to get continuity in second-term time, we should have to draw these lines so close together that no gaps could be noticed between them. The result would be a completely black block on which we should be unable to indicate the varying positions of *O*. There are two ways of dodging that difficulty. We can separate the horizontal lines thus,

```
G·············H    G·············H
G·············H     ·O
G·············H    ·O
G·············H   ·O
G·············H  ·O
G·············H ·O
G·············H·O
G············O H
G··········O·H
G·······O····H
G····O·······H
G··O·········H
Go············H
GO············H
```

or we may consider only a few points in *GH*, thus;

$$G \cdot \cdot \cdot \cdot \cdot \cdot \underset{O}{\cdot} \cdot \cdot \cdot \cdot \cdot \cdot H$$

and draw the past and future states of this row of dots above and below. Then, when we have drawn the vertical lines connecting the selected points in GH with the corresponding points above and below, and have represented the continuity of the experimenter's instruments by a diagonal line linking together all the *O*'s, we shall have a picture like this:

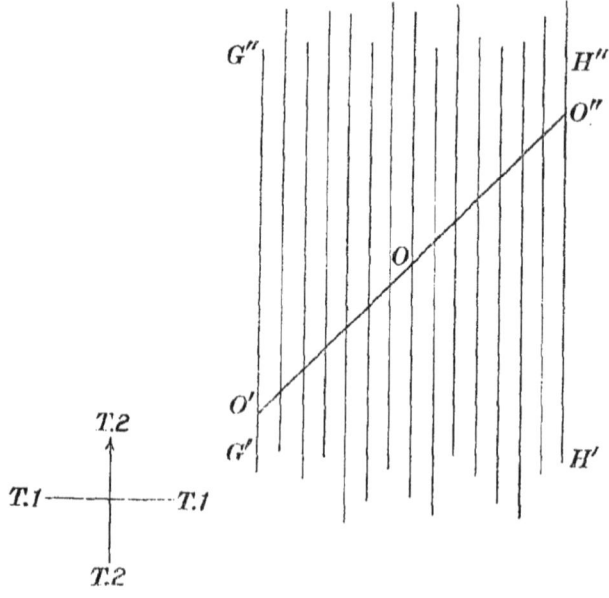

FIGURE 10.

The chief advantage of this diagram is that it throws into relief the *world-lines* which pertain to second-term time. I had best, perhaps, explain this at some length. The two little crossed lines drawn by the left-hand bottom corner of the figure serve very much the same purpose as does the little compass one finds printed in the corners of some maps. The compass shows which dimension of the map represents North and South, and which represents East and West. Our present little cross shows which dimension of our paper represents first-term time, and which represents second-term time. First-term time we shall refer to in future as 'Time 1' : second-term time we shall speak of as 'Time 2'. Time 1, which we had hoped, originally, to be able to treat as real, absolute time, has turned out to be merely a fourth dimension of space in which the original observed system is extended. Time 2, which takes into account the motion of the first-term instruments along the fourth dimension, we are regarding as absolute time; but we are representing it by the up-and-down dimension of the paper in *anticipation* of the step where we shall have to regard it, not as real time, but as a fifth dimension of space—which will happen when we take the motion of the second-term 'now-mark' into consideration. The line *O'O"* shows the positions which the experimenter's instruments (the first-term 'now-mark') occupy in the fourth dimension (the side-to-side dimension of the paper) at different instants of time 2. It is, thus, the *world-line* of those instruments in a time and space diagram where space is the fourth dimension and time is the fifth. (The three 'ordinary' dimensions of space are not indicated; but can be dealt with algebraically, if we wish to enlarge, unnecessarily, the task we have set ourselves in this chapter.)

GH—if we had filled in all the points along its length—would have extended into the past and future parts of time 2 as a 'world-*plane*'—thus making the black block on the paper which we are trying to avoid. The left-hand edge of that world-plane would have coincided with our present line *G'G"*. So *G'G"* represents the world-line, in fifth-dimensional time, of that point in *GH* which is *G*. Similarly *H'H"* represents the time 2 world-line of whatever configuration is represented by the point *H*. And the intermediate vertical lines in our figure represent the time 2 world-lines of those few points along *GH* which we have decided to take into consideration. All those points in *GH* represent cross-sections of a line (not drawn) which stretches along the fourth dimension (time 1). The positions of these cross-sections in that fourth dimension are fixed, and do not—like the position of *O*—change in the successive states of that line (unless the experimenter interferes). Consequently, the time 2 world-lines of these sections run straight up the paper parallel to the axis of time 2.

We have still to represent the time 2 'now-mark', which is the ultimate physical thing that we are considering—so far. We can do this by ruling a horizontal line *PP'* from *G* to *H* across the middle of the figure, and by adding an arrowhead to the time 2 line of the little dimension indicator.

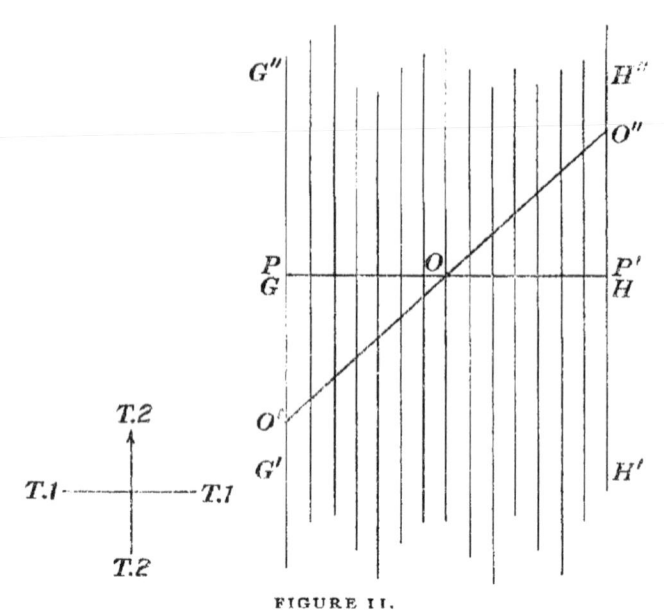

FIGURE II.
(FIGURE 9 in *An Experiment with Time*, first edition.)

It must be grasped that this diagram (representing the third-term world) consists of three parts. First, there is the original system which was objective to the experimenter's instruments. This was a three-dimensional world; but, in the analysis of the regress, it has expanded into a four-dimensional and, afterwards, a five-dimensional world. It ought to be represented by a plane *G'G" H"H*, but for convenience we have substituted for that plane a grid of vertical lines. This grid represents a time 2 map of the original object system. That system, no matter to how many dimensions it may prove to extend, we shall refer to, usually, as the 'substratum'. Upon the time 2 map of this substratum we have imposed a time 2 map, *O'O"*, of the first-term system of instruments. Then, upon this combined map we have imposed a line *PP'* representing the (so far) ultimate 'now-mark'; but we have drawn no *time map of the past or future states of that physical thing*. For the thing represented by the line *PP'* is travelling over the time 2 map. Consequently, the whole diagram is a 'working model', and real time is the time (not indicated) which times the movement of *PP*. This will be time 3. The time map of *PP'* showing the different positions of *PP'* at different instants of this real time, would need to be mapped out in a sixth dimension.

At *O* in the middle of FIGURE 11 there are three entities, viz., a point in the substratum, a point in the world-line *O'O"* of the first-term system of instruments, and a point in the (so far) ultimate 'now-mark' *PP'*. It will be more convenient in future to regard *O* itself as the intersection point of *PP'* and *O'O"*, rather than as one specific state of the instrument. It indicates, in this way, the *place* in the diagram which is 'now' in time 1. Clearly, it must travel up the

56

diagonal line $O'O''$ as PP' moves up the diagram. In travelling up $O'O''$ this point O travels, obviously, from left to right of the diagram, coming upon the original entropy configurations of the substratum (now represented by the vertical lines) one after another in order of that absolute time which is not yet pictured.

We are not yet in a position to describe our FIGURE 11 as a pictorial representation of the regress of observer and observed for which we are seeking. The entities shown in that diagram cannot be fitted yet into the table on page 30 —the table which we drafted thus:

World as observed by B_1	A_1			
World as observed by C	A_2	B_1		
World as observed by D	A_3	B_2	C	
				D

Nevertheless, we can prepare from FIGURE 11 a very similar table showing the 'now-marks' as geometrical determinants which abstract from a real world a series of worlds of progressively fewer dimensions terminating in the three-dimensional world of 'ordinary' space. Here it is: compare it with FIGURES 9 and 11.

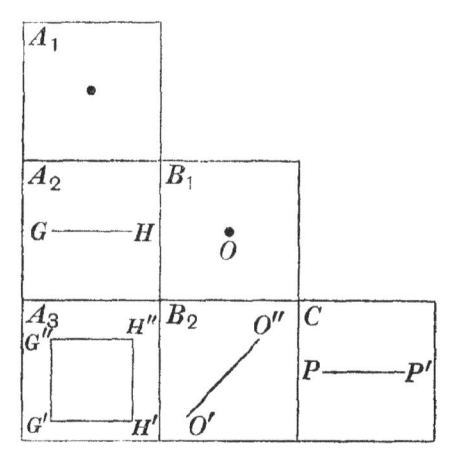

We read it as follows. B_1 is the first-term travelling 'now-mark'. It abstracts from the four-dimensional world A_2 (or GH), along which it is travelling, the three-dimensional world A_1. But B_1 (or O) is itself abstracted from the diagonal world-line B_2 (or $O'O''$) by the second-term 'now-mark' C (or PP') moving up time 2. And A_2 is abstracted from the five-dimensional world A_3 (or $G'G''H''H'$) by that same C (or PP').

Note that A_1, B_1 and C along the diagonal edge of the table, represent, respectively, the ultimate abstracted object, an abstractor 1 and an abstractor 2, as required by the table on page 29. And the curious feature which we noted in that table—viz., that B_2 does not abstract A_2 from A_3 is borne out in the present analysis: *0'0"* does not abstract *GH* from *G'G"H"H'*. Clearly, we are getting 'warm', and it may repay us to examine the nature of these 'abstractors'—these 'now-marks'—rather more closely.

CHAPTER XII

THE IMMORTAL OBSERVER AND HIS FUNCTIONS

An experiment is made, and the object system—the world external to the experimenter's instruments—is disturbed. It has received an impulse, and the physicist cannot account for its behaviour as subsequently observed unless he takes that impulse into consideration.

The problem of the origin of the impulse is one which the older philosophies enabled him to ignore. They assumed that it was possible to include both observer and observed in one and the same four-dimensional system, so that the classical laws of physical causation would suffice to account for every kind of physical interchange between the two parties concerned. Consequently, the physicist could leave the question of the origin of the impulse to the physiologist. The latter, however, could not start work until the physicist had laid down laws for his guidance. And the instructions which the physiologist received were simple. *He was not to take into account the possibility of any intrusion from any world outside the supposed single temporal system.*

But, if time in physics is regressive, those instructions no longer may be issued. And the physiologist is brought to a standstill. He must wait until the physicist can tell him *whence* he may regard that impulse as coming.

Now, whatever supplies the impulse must experience a reaction, and is, thus, an observer of that reaction. In this chapter, I propose to deal solely with that kind of observation which consists in recoil. With this proviso, I shall refer to the instrument as 'Observer 1', and shall speak of the ultimate source of the impulse as the 'Ultimate Observer'.

We saw in Chapter IX that the second-term 'now-mark' constitutes for the experimenter a *facility for moving the instrument*. We have represented that mark in FIGURE 11 by the line *PP'*. (This figure is reprinted on this page) It will be remembered that we are, for simplicity, regarding *O*, not as one specific state of the instrument, but as a mere abstraction—the intersection point between *PP'* and *0'0"*—the *place* in the diagram which is 'now' in time 1. And it has to be thought of as travelling in time 1, that is, as moving from left to right in the diagram.

Since *O*, while travelling in time 1, has to remain in *0'0"*, it must be considered as travelling up time 2 (the vertical dimension of the diagram).

Consequently, the physical thing which, at the point O, gives the impulse to the instrument is travelling up time 2. Since this thing is the recipient of the reaction, we may call it 'Observer 2'. This observer 2, then, has a *field of observation* travelling up time 2. But the thing which determines for this entity the order of succession in which the states of the instrument

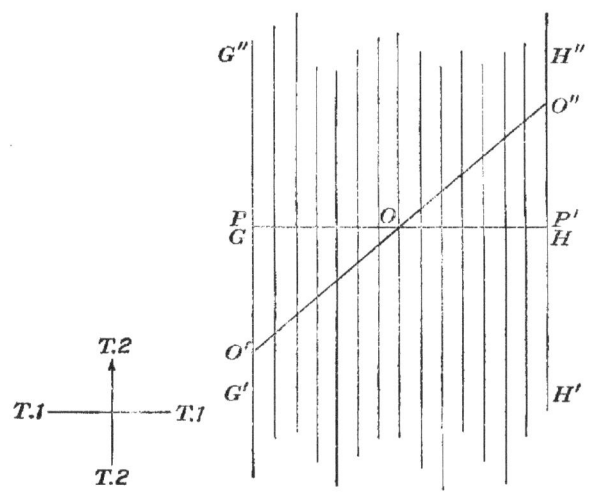

arranged along OO'' are presented for observation is—the time 2 'now'. So, for this observer 2, time 2 is real time. And this means that his own motion in time 2 must be *parallel to the time 2 axis*.

We might say, at once, that since time 2 is time for him, he belongs to the second-term world of four dimensions (with time as a fifth), and, so, is a four-dimensional entity. But we can give an additional argument for this. We have seen that his field of observation lies in the time 2 'now', and travels straight up the diagram. That field cannot be shorter in the fourth dimension than is the time 1 interval between his first and last observations of the instrument. He has observed this instrument at G' (when the time 2 'now' was there) ; he will observe it at H'' (when the time 2 'now' reaches that level) ; and his field of observation moves straight up the diagram during the interval of time 3 between these observations. That field must extend, therefore, in the present diagram, the whole width of the figure. He, therefore, is the second-term physical entity PP'.

Is it possible, now, for us to regard this observer 2 as that ultimate source of the impulse—the experimenter?

No.

FIGURE 11 shows the present state of the five-dimensional world; but it had past states, (when PP' was at the bottom of the diagram), and will have future states (when PP' will have moved to the top). If we drew diagrams of these states we should be showing past, present and future states of PP'—in time 3. We should discover then that the considerable disturbance we are visualising can be effected by the *experimenter* when PP' is in its present state, but not when it is in its past or future states. We should discover, also, that there is nothing in the physical characters of its past, present and future states to provide it with a unique ability for this interference at the time 3 'now'. So PP' does not represent the experimenter. It is a second-term physical instrument. And the time 3 'now' exhibits itself as a facility for altering that instrument. And so the argument goes on, *ad infinitum*.

The physicist introduces that multi-dimensional world and that endless series of physical instruments of more and more dimensions whenever he thinks of the object system as a series of states (or, for that matter, of events) in time, and as a system which can be made the object of experiment.

The non-technical reader may be inclined to wonder how it is that this observer 2, which is a four-dimensional thing with a four-dimensional outlook, can observe a three-dimensional thing like observer 1—the first-term instrument. There is, however, no difficulty about that, when the thing observed is resistance to force.*

He may wonder, also, what sort of a thing a four-dimensional instrument can be—from the physical point of view. But that aspect of the matter does not disturb modern physicists, most of whose work is concerned with four-dimensional entities. Some of them would declare that the four-dimensional substratum GH consists of a recognised physical quantity known as *Action*. This has the dimensions of Energy multiplied by Time, and we shall have a great deal to say about it later on. For the moment, it suffices to point out that PP' is an entity of exactly the same physical dimensions as GH.

But that brings us to a really unexpected fact. The regress compels us to regard PP' as a real entity abstracting an unreal O from a real $O'O''$ (vide the table at the end of the last chapter). Moreover, a body which you are employing for the observation of a second body does not become unobservable whenever that second body is absent. Consider, then, what happens to the entity PP' when it is not utilising the first-term instrument at O—consider, for instance, that this first-term instrument gets broken and, subsequently, is repaired. (We should show that state of affairs in FIGURE 11 by making a gap in the middle of the line $O'O''$.) While passing over that gap, PP' would continue to exist, ready to re-commence observing an O as soon as the gap in $O'O''$ had been traversed. Now, the truth of that assertion would not depend upon whether the gap in $O'O''$ were long or short. Clearly, then, its truth would not be affected if the instrument were never repaired.

Would this continued existence of PP' in time 3 be affected if PP' did not extend beyond the left and right-hand edges of the diagram? (The substratum itself extends, of course, a long way in both directions beyond those limits.) The answer here, again, must be in the negative.

Turn now to the substratum. In the time 1 dimension (from left to right) its character is differentiated; i.e., each state represented by a vertical line is different from the state next to right or left. But there is no differentiation in the

---

* Note for physicists: It must be remembered that time, for this observer 2, is the fifth dimension.

vertical dimension, *above GH*. Such differentiation would be logically impossible. For the states from left to right are supposed to be related to one another in the manner dictated by the laws of classical science : they represent a causal scheme. An interference at *0*, for example, (see FIGURE 12), would change all the states between *0* and *H*. The vertical lines above *OH* would become then different from all the lines below *OH*, but that breach in the continuity of the lines cannot occur at a level not yet reached by *PP'*. Suppose, for instance, that the line running up from *0* had a changed character above a point *Q* a little ahead of *PP*. All the lines above *QR* at that level would have correspondingly changed characters, so that a causal relation could be traced from *Q* to *R*. But below *QR* the lines would be causally related so as to agree with the different condition of the line between *0* and *Q*. Then, as *PP'* moved upward, observer 1, travelling from *0* to *S*, would come upon substratum states in a certain causally related condition. (We are ignoring microscopic physics in this chapter.) But on arriving at *S*, he would encounter a state belonging to an entirely different causal scheme originating at *Q*. He would discover that a miracle bad happened !

Now let us consider that *PP'* has travelled up the diagram as far as the level *VO''*, and that, at this level, the instrument *0'0''* ends through, say, breakage. Alternatively, let us say that *PP'* extends no farther than the width of the diagram. In either case, when *PP'* reaches *VO''*, its chance of

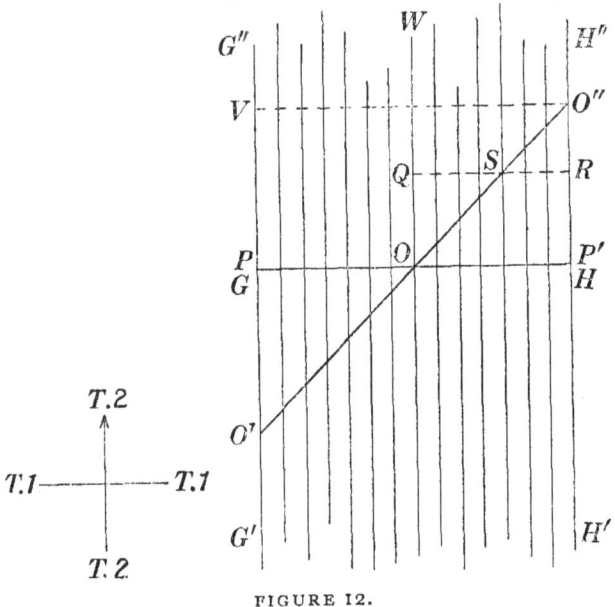

FIGURE 12.

interfering with the substratum ends. For it is our initial supposition that the experimenter can interact with the substratum at the time 1 'now' only; that is to say, can interact only via the first instrument. (Interaction at any other point would produce miracles for the observer at *0*.) *Hence*, after PP' has passed VO'' neither the substratum nor PP' can effect changes in each other; the lines above VO'' persist unaltered in time 2 for ever; and PP' moves over them in time 3.

What, then, is to interrupt the continued existence (in time 3) of this observer 2? Nothing save a miracle.

Now, PP' is not the experimenter: it is one of that individual's instruments. It, like the first-term instrument, is one of an endless series of 'observers' intervening between the experimenter and the substratum. And the really interesting thing is the way in which those observers are related by the time device.

Everything in the diagram which runs from left to right is differentiated in that dimension. The result of that differentiation is, as we all know, a beginning and an end in time 1 for any entity which depends for its identity upon a condition of internal organisation. Let us assume, for security in this vital question, that everything pertaining to the experimenter is limited in this way, and let us say that the width of the diagram indicates those limits.

Observer 2, as we have seen, will lose touch with observer 1, leaving it behind him in the fifth dimension. A moment's consideration shows us that this is simply because observer 1's world-line $O'O''$ crosses the diagram from left to right, that is to say, from beginning to end in time 1. But observer 2 thereafter travels straight up between those two boundaries, and there are no limits or changes assigned to the substratum ahead of him in time 2, and no limits assigned, as we have seen, to his endurance in time 3. The endurance of the substratum in time 3 would have to be shown by arranging a series of diagrams like FIGURE 12 one above the other in the fashion of the leaves of a book, making a tall pile.* The pile would have boundaries on the left and the right, *but no boundaries towards the tops and bottoms of the pages. And it would be unlimited in height.* The successive positions of PP' in that pile, each a little more towards the top of the page than the one next below it, would build up an inclined plane endlessly long in the time 2 and time 3 directions. Observer 3 would be represented by a horizontal level taken through the pile. It would form a plane with boundaries on the right and the left but none in the time 2 direction. Its travel would be a rising motion up the tall pile, that is to say, in the time 3 direction. A little consideration shows that it would never lose contact with observer 2 (the inclined plane). Also there would be no limits to its endurance in time 3 (the time which times its travel).

In brief, of the entire series of observers, the only one which comes to an end in its own time dimension is observer 1.

But observer 2 cannot interfere with observer 1, after he (observer 2)

---

* See Appendix for a perspective drawing of this figure, taken from *An Experiment with Time*.

has passed the line *VO"*. What about observer 3? He can continue to interfere with observer 2; but he cannot interfere with observer 1 except via observer 2, so, when observer 2 loses touch with observer 1, observer 3 is rendered impotent to interfere with observer 1. And the same restrictions apply to all the other observers.

The first-term 'now' at *O*, the intersection point between *PP'* and *O'O"*, represents the experimenter's chance of altering the substratum ahead of *O* in time 1. Such alteration changes, as we have seen, that part of the substratum which is ahead of *both O* and *PP'*, viz., the rectangle *O WH"P'* in FIGURE 12. This alteration changes that corner of our imagined pile which lies ahead of *O* and of *PP'* and of the third-term 'now-mark'. And so on throughout the series. Thus, interference at *O* alters what lies ahead in the time pertaining to every observer in the series. But, once observer 2 has passed the point where observer 1 intersects the right-hand boundary of the diagram, *O* vanishes, and the experimenter has lost his last chance of altering the future course of his originally selected object world.

The observational powers of observer 2 in the absence of observer 1 are matters of great importance to mankind, and we had best look into this question very closely. We have proved that this observer, *PP'*, is travelling parallel to the time 2 axis and is possessed of a field of observation extending at least from *G* to *H*. We have seen that he continues to exist in the absence of observer 1, e.g., when observer 1 is inactive, or when observer 2 has passed the position *VO"* (FIGURE 12). But does this mean that *PP'* observes *the substratum,* and continues to do so when he has no first-term observing instrument to assist him? The answer is in the affirmative.

First, I may repeat here the argument given already in *An Experiment with Time* (3rd edition, pp. 179-181).

'The development of the series of observers places observer 1 (the section of *O'O"* which is at *O) between* observer 2 and the substratum section at *O* which is, somehow, affecting that observer 2. So that the process by which that particular state affects observer 2 is as follows. A certain feature in that state causes a corresponding modification in the intervening section of *O'O"*. It is this reproduced feature which affects observer 2.

'But that raises the following difficulty. Observer 2 is a four-dimensional creature, and the section of *O'O"* which intervenes between him and the substratum is only three-dimensional. His field of observation must extend, therefore, in the fourth dimension beyond the place where *O'O"* crosses that field. In those outer parts of observer 2's field there are many other three-dimensional sections of the substratum containing the kind of feature which, reproduced in the intervening entity, is affecting observer 2. Since observer 2 is susceptible to features of that kind, what is there to prevent him from being affected by these other three-dimensional sections of the substratum as well as by the section of $O'O''$ which lies in his field?

'Nothing, that I can see. So, pending the discovery of some obstacle, we must assume that observer 2 is affected by the substratum adjacent to the section of $O'O''$. *But this collection of adjacent sections does not affect him in the same way that he is affected by the three-dimensional section of $O'O''$.* The bit of the substratum beside $O'O''$ is a four-dimensional strip presented as a *whole* to a four-dimensional observer—it has, to him, no distinguishable three-dimensional sections. The function of observer 1 (i.e., the function of the only purely three-dimensional entity within the field) is to abstract from the substratum an aspect thereof with which, otherwise, observer 2 could never become acquainted.'

All of which is reasoning sound enough.

But in the present book we can arrive at the same conclusion by a simpler route. As we saw in the table near the end of Chapter XI, PP' (or C) is a geometrical abstractor abstracting GH (or A_2) from $G''H''H'G'$ (or A_3). We can add to this what we have proved earlier in the present chapter —that he is an observer with a field of observation as long as GH. Therefore he is an *observer* abstracting GH from $G''H''H'G'$ (an A_2 from an A_1). Clearly, then, he is the observer C of the table of the self-conscious observer on pp. 29, 30 of Chapter VI. And O, the part of $O'O''$ which is 'now', is his first-term 'self'.

But how, the reader may ask, can this PP' observe the substratum when (in the absence of observer 1) he is not *being altered* by it? For, after the disappearance of observer 1, PP' simply rushes on over a substratum which never alters and to which he is already, so to say, 'fitted'! Here I must refer the questioner to the definition of physical observation given in the first page of Chapter V. To 'observe' is to *'be-affected-by'* and not necessarily to be *'altered-by'*. Suppose that $PP's$ form is adapted to the form of GH, so that in the absence of GH, $PP's$ form might be otherwise. That would mean that $PP's$ freedom is being *restricted by* the presence of GH. And *that* would amount to physical observation. Our proof that PP' does observe GH has led us, therefore, to no physical absurdity.

It is to be noted that this regress of time clears up the difficulty we discerned in our general table of a self-conscious observer. PP' can perceive perfectly well that the substratum is altering O as O travels (in $PP's$ field) from left to right across it. For PP' can see any point in the substratum ahead of O in his field, and can notice that O is changed to conform with the new conditions when it arrives there.

THE SERIAL UNIVERSE

PART III
SPECIAL TESTS OF THE THEORY

CHAPTER XIII

AN APPROACH TO RELATIVITY

How fast does the 'now' travel?

At first sight this question seems either meaningless or impossible to answer. Very well, let us see what a second inspection will make of it.

To begin with, we must realise that the question is not one of deciding how fast our instrument at O (the B_1 of our table) is travelling over an *already marked out* space and time map (A_2 in the table), i.e., how fast O in FIGURE 12 is travelling over an already marked out GH. Our problem is to employ the knowledge of the object world provided by our instrument B_1 for the purpose of constructing precisely such a marked out A_2 map—and this when we have not the faintest notion of the rate at which that instrument is travelling over the fourth-dimensional length of the countryside to be plotted out.

Well, let us start with the part of our task which is easiest. Our set of instruments in the B_1 system contains a scale of distances in ordinary three-dimensional space, and the travel of that scale in the fourth dimension will not alter its length. We can employ that scale, therefore, to mark out a scale on the space axis of our map.

But trouble arises when we try to mark out a time 1 scale, keeping in mind that it indicates a fourth dimension of space along which our source of information, B_1, is travelling. For our clock is not something which B_1 (our set of instruments) observes : it is something which *we* observe without the intervention of instruments. Its ticks are not features in a time 1 *world-line,* but events which we have to mark out on that time 1 *axis* to which time 1 world-lines will be referred. It belongs, in brief, to the system of present, three-dimensional instruments which provide us with the information from which we propose to draft a time map of the endurances of bodies other than that clock. That is to say, the clock belongs to the travelling B_1.

Probably, this will be grasped more easily by the employment of a diagram. Let us assume that our clock's ticks occur at intervals of one second. FIGURE 13 shows axes of time 1 and time 2 (indicated by T_1 and T_2). The world-line of our B_1 clock (a B_2) may be represented by any sloping line we please, such as OO''. Its ticks will be features in its career—features which we can represent by marks made at regular intervals along its length. These ticks are, thus, periodic in time 2; and we can use them to mark off a scale of T_2 seconds, by drawing horizontals to the T_2 axis. But the ticks are periodic also in time 1; so they will serve to mark off a scale of T_1 seconds, by dropping verticals on to the T_1 axis. Clearly, at whatever

angle we draw OO'', the diagram will indicate always that the clock is travelling along time 1 at the rate of one time 1 second per second of time 2. In other words,
$$t_1/t_2 = 1, \text{ i.e., } t_1 = t_2$$
One what? To give this velocity a meaning we must realise that the seconds marked off on the T_1 axis are marked by the ticks of a clock travelling

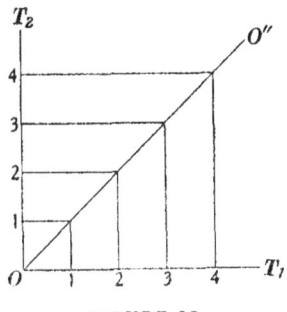

FIGURE 13.

along a fourth dimension of *space*, so that the T_1 'second' represents a *space* distance travelled by the clock in one second of time 2. The clock is then marking off both real seconds in time 2, and space lengths in the fourth dimension. And the time observed by us—the time told by the clock—will be time 2.

Now, we are all agreed that the rate at which a clock hand travels over its dial must be assumed to be constant if that clock is going to be accepted as our measure of time. Therefore, since our clock ticks out the seconds of time 2 *(vide* FIGURE 13) we must not only regard its ticks as evenly spaced along OO'', but must regard also the line OO'' itself as straight, i.e., our clock is travelling along the fourth dimension at a uniform velocity. If we call the fourth-dimensional space-length traversed in any period of time 2, S_4, then the velocity of the clock will be

$S_4/t_2 =$ a constant.

We will call this constant, k. Then

$t_1/t_2 = k$, a velocity.

And $\qquad\qquad t_1 = kt_2 = S_4$ * $\qquad\qquad$ ......(2).

Now, k, the velocity, means

$$\frac{k \text{ units of space 4}}{\text{one second of time 2'}}$$

where k is a mere number, and the distance travelled in one second of time 2 *is (vide* FIGURE 13) one division of our intended time 1 scale. So that, in preparing our four-dimensional map, we shall have to give each second-division of time 1 the same length as we give to k units of three-dimensional space.

---

* See note on the following page.

And there we stick. What number of space units is k? We have not the faintest notion. And it is obvious that we shall never discover its value so long as we continue on our present lines.

Let us try another method.

Note. Readers who are unacquainted with equations of the kind we have been considering may be momentarily puzzled by the assertion that $t_1 = kt_2$, when we have seen, a little earlier, that $t_1 = t_2$. They might suppose even that k must be the number 1. But t is really an abbreviation for t *[T]*, an expression in which *[T]* is the *unit* of time (i.e., one second) and t *is* a mere number. Similarly, s is an abbreviation for s *[S]*, where s is a pure number and *[S]* is the unit of space. The unit of velocity is *[S] / [T]* and the velocity k means $k \times$ *[S] / [T]*, where k *is* a pure number. Hence, kt_2 means $k \times$ *[S] / [T₂]* $\times t_2$ *[T₂]*. Since the two *[T₂]*'s cancel each other, the expression resolves itself into kt_2 (both pure numbers) units of space. Multiplying the time t_2 by the *velocity* k does not, therefore, alter its magnitude: it merely expresses the t_1 magnitude as being equivalent to kt_2 units of space.

CHAPTER XIV

VELOCITY OF THE 'NOW'

We have arrived at the facts that each second of our time 1 scale will have to be made equal to a distance of k space units, and *must represent also the distance which would be traversed by the travelling clock while it ticked one second of time* 2. But, for our simple scale to represent these facts, it was necessary for us to assume that the seconds of time 2 were being regarded as marks on an axis drawn in a fifth dimension, and that the world-line of the clock was being represented by the inclined line in FIGURE *13*. Failing that or some alternative understanding, our time 1 scale, with its divisions of k space units apiece, would represent nothing but a space length over which anything might be travelling at any rate whatsoever. Now, our method of showing that the divisions represented the distances traversed during clock ticks was perfectly sound. But there is another way of making our scale show what is required of it; and, since we have been brought to a standstill, we had better see what this other way will do for us.

A quantity which can be represented diagrammatically by the length of a line—that is to say, by some marked-off distance on a scale—is called a *'Scalar'*. Ordinary space length and ordinary time duration are examples of simple scalars.

But now we have to consider quantities of another kind—quantities which can be represented by lines of definite lengths and fitted with arrowheads. Such lines are called *'Vectors'*, and they specify several aspects of the quantity they represent. That quantity is to be conceived as a *transportation* or *transference* or *step* from one end of the line to the other. To quote A. N. Whitehead : 'All other types of physical vectors are really reducible in some way or another to this single type'. The arrowhead gives the *sense* of the transportation, i.e., tells us from which end to which end of the line the transportation is supposed to be taking place. If we place the line within the angle made by two axes, the slope of the line will give what we may call the orientation of the transportation. The ends of the vector will tell us where (as referred to these axes) the transportation starts and ends. The length of the line indicates the *amount* of the transportation. This amount is a simple scalar quantity, and it can be indicated either by a scale marked on the line or by referring the line to scales marked on the aforesaid rectangular axes.

If we announce that this amount of transportation is to be considered as the distance moved in a constant interval of time, then the length specifies a velocity, and a long line will represent a larger velocity than is indicated by a short one. There are other quantities which the length of the line may be made to represent (by suitable conventions) but we need not stop to consider these.

The point to be borne in mind is that every vector possesses, besides its other characters, its character as a *scalar,* which is the character represented by its length. To distinguish this scalar character of a vector from scalar quantities represented by lines which are not vectors, we call the former, a *'Tensor'.* A tensor is simply the scalar belonging to a vector.[*]

Let us turn now to the scale we want to mark off along the fourth-dimensional axis of the mesh-system pertaining to our intended four-dimensional map. Each unit interval thereof will possess, as we saw in Chapter XIII, a length equal to k space units—and we do not know the value of k. Of course, if we could ascertain the length of one of these interspaces, we could use that length to mark out all the remainder. But the only way in which we can discover that length is (as we saw in the last chapter) by discovering first the unknown velocity k with which our clock is travelling in the fourth dimension, and by marking thereafter the places it has reached in that dimension at the beginning and end of one of the seconds it is ticking out in time 2. From the formula $kt_2 = S_4$ we could then calculate the length of the distance thus marked out.

But the little line arrived at in this fashion will have a very curious character. Indeed, it will have two distinct characters. In the first place, it

---

[*] 'Tensor', employed in this sense, is a word pertaining to a certain classical calculus which is going to prove of great importance to our problem. 'Tensor', in the vocabulary of modern relativity, has a different meaning.

will indicate a unit distance travelled by the moving clock. In this capacity, it is a *unit vector of transportation with its tensor (scalar value) measuring S_4*.

In the second place, it will be a pure scalar measuring a second of time 2 —since it is marked off by the ticks of the uniformly travelling clock. It specifies, in fact, *both* a distance travelled and the time taken in travelling that distance.

We have come upon a length of that kind in our everyday life. An interval marked upon the circumference of a clock can have that double character. It can specify both the amount of a displacement of the clock hand and the time in which that displacement is effected.

Let us remind ourselves once more of what we are doing. We have a three-dimensional clock which, according to serialism, is travelling along S_4 and ticking out seconds of time 2 (time 1 is S_4). Say that at two successive ticks we observe two objective features in the fourth-dimensional path over which our instrument is travelling. We want to mark upon our S_4 scale the fourth-dimensional distance between those two features. And we realise that, if we succeed in doing this, the distance marked will be both a distance moved-over in S_4 or T_1 and *an interval upon a scale of T_2*.

What we had hoped to do was to make the interval an S_4 length only, and then to bring in time 2 as a fifth dimension at right angles to that length, as in FIGURE 13. We had intended, thereafter, to draw a diagonal line between the axes of S_4 (or T_1) and T_2, which line should indicate, by reference to those axes, the rate at which the clock was travelling. Then we should have indicated the travel of the second-term 'now' by an arrow pointing up time 2. (That, of course, would have made the T_2 lengths represent amounts of transportation.) We shall be able to draw something like (though not exactly like) that picture—after we have discovered the sought-for velocity. But, to find that velocity, we are obliged to draw, first, a picture in which the scale of time 2 and the scale of S_4 (or time 1) occupy one and the same position. Now, we can show diagrammatically exactly what it is we are doing when we draw this preliminary picture. FIGURE 14 shows the picture we want to arrive at, with the arrow pointing up time 2, and with the unit of time 2 marked on the T_2 axis. The unit of time 1 is shown as a space length S_4. If, now, we were to *rotate* the vector of T_2 *with its arrow*, about the pivot point O, until it lay along the axis of S_4, with its T_2 divisions coinciding with the S_4 divisions, and its arrow pointing along S_4, then we should have drawn exactly the picture which is presented to us by our method of exploring S_4 with an instrument travelling in that dimension *(vide* FIGURE 15).

FIGURE 14.

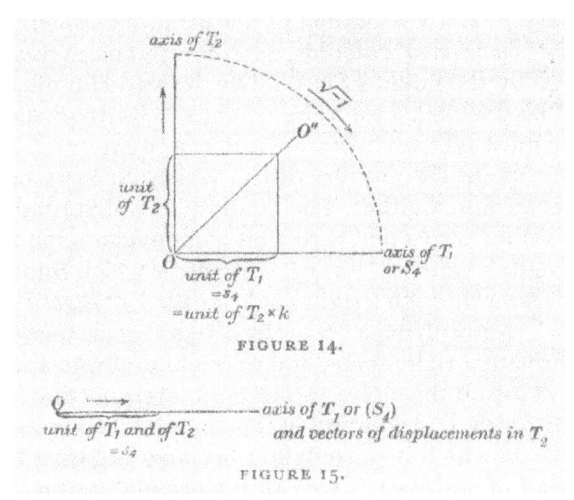

FIGURE 15.

For here the horizontal line is the tensor (scalar character) of a vector of transportation with the necessary arrow, and is also a scalar indicating the unit of T_2, i.e., the time recorded by the clock.

Now, how are we to shut up FIGURE 14., concertina-fashion, until it presents to us that FIGURE 15 which is the only picture we can draw when we explore the four-dimensional world with our travelling instrumental system?

There is only one way to effect this, and that is to multiply the unit of T_2 by the square root of minus one (written $\sqrt{-1}$).

Why? I am sorry, but, to see why, the reader will have to study a branch of mathematics known as the '*Quaternion*' calculus, and invented many years ago by the famous Sir William Rowan Hamilton. If he does not wish to be troubled with that, then he must take my word for it that a 'quaternion' is the name for any operation which changes one vector into another. The quaternion which rotates a vector into a new direction without changing its length is called a '*Versor*'. The versor which rotates a vector through a right angle is called a *'Right Versor'*. Multiplying a vector by a right versor turns it through a right angle, and a second multiplication will turn it through another right angle; so that, at the finish, it is pointing in the opposite direction to that in which it started, and becomes negative instead of positive. If we call the original vector, β, and the right versor, i, the total operation amounts to $i \times i \times \beta = i^2\beta = -\beta$, whence $i^2 = -1$. So that i, which, when multiplied by P, turns that vector through one right angle, equals $\sqrt{-1}$.

Consequently, if we multiply the vector T_2 in FIGURE 14 by $\sqrt{-1}$, we shall rotate it through a right angle into its new position in FIGURE 15. But then the value of the S_4 length (the T_1 unit) will be, not $k \times$ unit of T_2, but $\sqrt{-1} \times k \times$ unit of T_3. For the vector in T_2, the vector which is rotated, is the velocity k *up* T_2 multiplied by the time which is not indicated in

FIGURE 14—and *that* time is the merely scalar quantity T_3.

This gives us our clue. Glancing at FIGURE 14, and remembering that the 'now 1' is the intersection point between the 'now 2' and the diagonal *00"*, we see that the speed *k* of the 'now 1' along S_4 is the same as the speed *k* of the 'now 2' up T_2. We want to express the former speed as length of S_4 traversed per ordinary scalar unit of a time 2 which is not in the proposed diagram. Clearly, a diagram of ordinary space *S* and of T_1 can be treated in the same way as we dealt with the diagram (FIGURE 14) of T_1 and T_2. That is to say, we can multiply the T_1 vector by $\sqrt{-1}$ and rotate it so that it lies along our axis of ordinary space. This will give us a length of ordinary space *S* equal to $\sqrt{-1}$. *k* x the unit of T_2, where *k* is T_1 / T_2.

Now, a classical physicist has no time vector which requires rotating, and, therefore, he has no need to multiply recorded time by $\sqrt{-1}$. Let us suppose, then, that he has to consider, in his four-dimensional map, an inclined world-line such as *ab* in FIGURE 16. Drawing from *b* the line *bc* parallel to the space axis, and from *a* the line *ac* parallel to the time axis, he would produce a right-angled triangle. He would express the length of *ac* as *T*, and the length of *cb* as *S*. We know that, in this right-angled triangle, the square on *ab* is equal to the sum of the squares on *ac* and *cb*. For him, then, the length of *ab* (let us call it, the 'distance') would be given by the following formula :

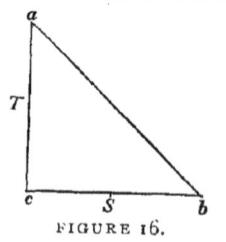

FIGURE 16.

$$T^2 + S^2 = \text{distance}^2.$$

But a serialist, who recognises that time 1 is a vector $kT_2 = S_4 / T_2 \times T_2 = S_4$, has to rotate that time 1 vector by the use of $\sqrt{-1}$ in order to get a proper description of the lengths in *S*. These lengths become equal to $\sqrt{-1} \times S_4$. His formula, then, would be

$$(kT_2)^2 + (\sqrt{-1} \times S_4)^2$$
$$= k^2 T^2 - S_4^2$$
$$= k^2 T_2^2 - S^2 = \text{distance} \dots\dots\dots(3).$$

The map constructed according to this rule—the map arrived at by watching instruments which are *travelling* along the fourth dimension—will show the four-dimensional hyperbolic world of Relativity.

Let us consider now the foot-rule with which our travelling instrument is equipped. We can represent this by the dotted line $B_1 B_1'$ in *FIGURE 17*. It is travelling along the axis of S_4 at the still undiscovered velocity *k*. The axis *S* is an axis of one of the other three dimensions of space. $B_1 B_1'$ is supposed to be intersected at the point 0 by a fixed world-line *ab* crossing $B_1 B_1'$ at an angle of 45°- (This assumes that we draw our mesh-system with the horizontal intervals equal to the vertical intervals.) As $B_1 B_1'$ moves with velocity *k*, 0 *will* travel down $B_1 B_1'$ towards B_1 with a velocity equal to *k*. Now, the velocity represented by the inclination of the above world-line *ab* at 45° in any world where the observed time is multiplied by this velocity *k*, and ordinary space is multiplied by √-1,

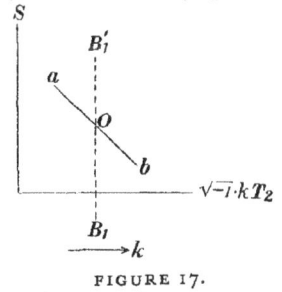

FIGURE 17.

will possess most extraordinary characteristics. This was proved by Minkowski. It will be a limiting velocity, inasmuch as nothing used for a *signal will* be able to travel at a higher speed. And any object which is travelling with that velocity will appear, according to the measurements of the three-dimensional observer, to shrink to nothing in the direction in which it is moving, while retaining its usual magnitude in the other directions.

It is this velocity which will appear to our travelling instrument as the velocity of 0 down the scale $B_l B_l'$—the velocity which is equal to *k*.

Here is a chance to see whether our serialism is right! Let us examine the universe around us with our three-dimensional instruments and see if we can find anywhere a velocity, in three-dimensional space, which possesses the above paradoxical characteristics. If we are lucky enough to discover it, it will prove that our method of assuming our instrument to be travelling in S_4 is right. For the magical qualities of that velocity will *depend upon* the travel of our instruments. And, incidentally, its velocity *k* in three-dimensional space, which velocity we shall be able to measure with our three-dimensional instruments, will be equal to the velocity of the 'now'.

We find it at once. It is the velocity of *light*. And it is known to physicists as the constant, *c*.

Our *k*, then, is this *c*, a velocity of 300,000 kilometres per second. And we can draw the meshes of our required mesh-system thus,

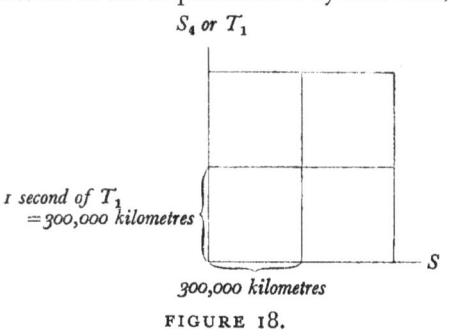

FIGURE 18.

The relativists did not proceed as we have done. Einstein began by *assuming* that light possessed irrational properties (in order to account for the results of the Michelson-Morley experiment). Minkowski discovered thereafter that, if time were regarded as a fourth dimension with units equal to c x the unit observed time, and if space were regarded as observed space x $\sqrt{-1}$, then the magic would be transferred from the behaviour of light to the unknown spell which changes observed time and observed space into two such extraordinary quantities.

We have shown that, if the regressive character of time is taken into account, the map of the world made by instruments which are 'now' must show a world in which observed velocities have an upper limit, and where a velocity at or near that upper limit will behave as light, quite rationally, does behave.

NOTE. It is true that our rotation of the vector CT_2 was done merely in order to discover the numerical value of C, and that we have returned it to its normal position in the space-time map FIGURE 17. But there is no need actually to change the normal direction of that vector in drawing the final map. In FIGURE 17 there are only two *entities* regarded as moving: the foot-rule and the clock in $B_1 B_1'$ travelling in the T_1 direction. The motion which B_1 measures as proceeding along his footrule is the motion of merely an *intersection point O* between the real foot-rule in $B_1 B_1'$ and the real, stationary world-line *ab*. Obviously, we cannot have in this diagram vectors of displacement in *both* S_4 and S—as if we had two *real* things moving at the same rate in different directions. A glance at the diagram suffices to show that the vectors of displacement of O in the S direction are only *records* of the real displacement of the foot-rule in the S_4 direction. Consequently, in a space-axis marked with intervals obtained from these records, the conditions *are* exactly similar to those which would be found if the vectors of displacement of the foot-rule *were to be* rotated. That is the meaning of saying that displacements in S are $\sqrt{-1}$ times those in S_4.

CHAPTER XV

THE REGRESS IN RELATIVITY

From now onward the reader will need to refer continually both to the table on page 57 and to FIGURE 11. I know from experience that it is most troublesome to have to hunt back for these two illustrations. Fortunately, it is just possible, I find, to print both on one page (*this page in this new edition*); and I have asked Mr Lewis to repeat the pair thus on the right-hand page which follows next. Then, if the reader slips a book marker in at that place, he will be able to make his references without difficulty.

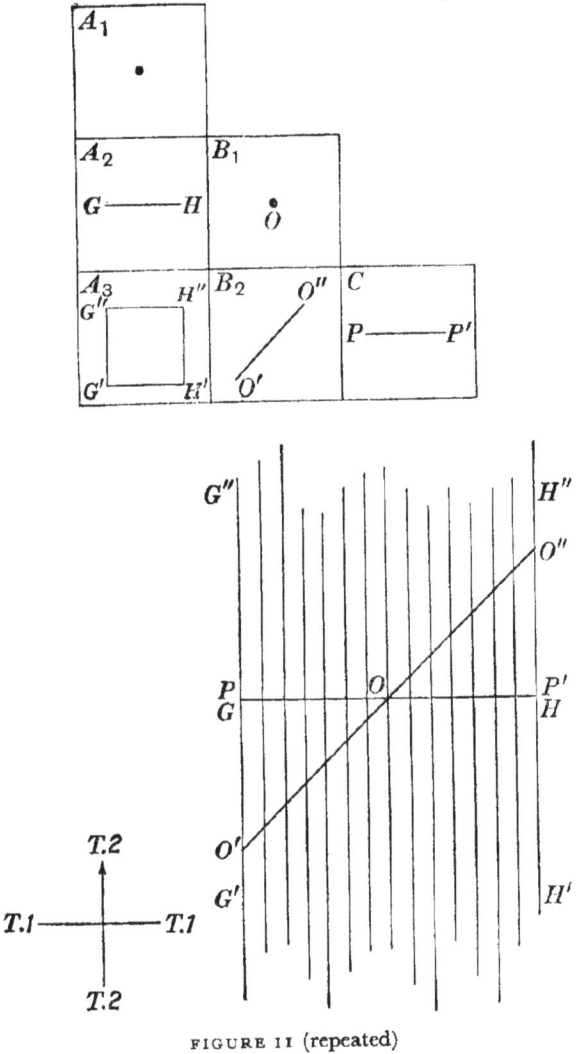

FIGURE 11 (repeated)

Glancing, then, at the table, he will see that the map we have just sketched out is a picture of the world as this would be observed by an imagined four-dimensional observer C_1 the observer who can see that the B_1 instrument is *travelling* along A_2 (or S_4). Now, this imagined observer can perceive that the travelling of B_1 along S_4 is taking time. The question arises, therefore, why he should not construct a map with time as a fifth dimension—a map which would show the different positions (in S_4) of B_1 at different instants of this fifth-dimensional time. But, as soon as we ask ourselves this, we find that the fourth-dimensional axis we have drawn does show the positions (in S_4) of B_1 at different instants of time 2; for the divisions on the scale of that axis indicate *both* the distance travelled and the time 2 taken in travelling that distance. Have we, then, got rid of time—stopped the regress?

Well, let us look at this S_4 axis again. Time 1 is marked out there, and so is time 2. Precisely, and this means that the instrument is travelling over time 1 *(S_4) but only marking out the past moments of* time 2. It is making a record, which is not the same as travelling over a record made. Time 2 is timing the travel over time 1. But can we say that the time 2 divisions represent the time taken in travelling over the time 2 divisions?

Perhaps the reader will think that this is hairsplitting. Surely (he might argue) we can say that time 2 and time 1 have become, now, one and the same absolute time, so that if the " now 1" travels over time 1 (which he agrees to) it is also the " now 2 " travelling over time 2.

But that would be untrue even *if* we could actually eliminate our T_2 dimension by the device *of* multiplying vectors therein by $\sqrt{-1}$ so as to lay them along T_1 in FIGURE 15. (Remember that we have not done this in our present map *of* space and time 1.)

FIGURE 14 is repeated on p.77. Note what it represents *before* the rotation takes place. Observer 2 (C_1) is travelling up the time 2 dimension, which becomes, consequently, a fifth dimension of space and is equipped with an arrow to show that its lengths represent vectors of transportation. But the time which is timing the motion of observer 2 along that axis is not time 2. *It is time 3*, which would be mapped out as a dimension of length in the next stage of the regress. And we cannot say that the T_2, or S_5, axis represents time 3 lengths unless we have multiplied, previously, the vector of T_3 by $\sqrt{-1}$, so as to rotate it into the position of the T_2 axis. The time 3 axis remains, consequently, sticking up above any four-dimensional map in which we claimed to have rotated the *axis* of T_2.[*] And it is *in this* fifth dimension that we should have to indicate the time taken by B_1 in travelling along the S_4 axis.

---

[*] Imagine the T_3 axis as standing out at right angles to the page on which FIGURE 11 is printed. Then imagine yourself looking up the diagram with your eye level with the bottom of the page, so that the whole figure is foreshortened into a horizontal line—but with *0 moving along* this. The T_3 axis will be then the axis of the diagram in which you have to plot out the successive positions of O.

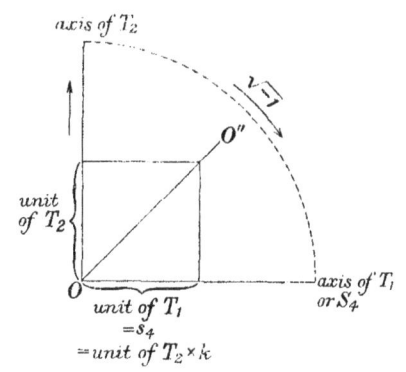

We have had to reach this conclusion by a rather tortuous route; but, now that we know where we stand, we can see that there is a simpler way of treating the whole matter.

Spread this book, open at this place, flat on a table. Raise the left-hand half vertically. Turn up page 77 till it divides the angle between pages 76 and 79. Then the near edge of page 76 can represent the axis of T_2 (or $\sqrt{-1}$. T_3), the near edge of page 79 can represent T_1, and the near edge of page 77 can represent OO''. The line on which these three pages hinge will serve for a dimension of ordinary space S. Page 77 will represent then the world-line of a B_1 instrument which possesses a space length equal to the length of the page-hinge.

If the sloping page 77 were out of the way, you could stand up a playing card with its left edge touching the near edge of page 76 and with its bottom edge resting on page 79 along a line drawn at 45° from the left-hand bottom corner of that page. The bottom edge of the card represents now the world-line of a particle—a world-line mapped out in S and T_1. The whole card represents the endurance of that line in T_2. If you want to erect the card without first tearing out page 77, you will have to cut a slit in the latter along the line where the card intersects it. This slit represents the succession of points where B_1 will find the particle at different instants of T_2. You can see this if you think of the book as standing in a pool of water the surface of which is level with page 79, and then imagine that water-level rising. The level itself represents observer C, and the line where it cuts page 77 represents observer B_1. As the level rises, B_1 moves to the right and observes a point in the slit as travelling steadily in the space dimension.

Next, imagine an electric torch shining through the slit from the right, so as to throw an inclined line of light on page 76. This line will run from the lower near corner of the page upwards and away from you at an angle of 45° to the space axis. It will show you what are the measurements of the slit as these are referred to the axes of S and T_2. As the water-level rises, this will mark off in that line also a point travelling steadily in the space dimension. The rate of travel of that point in that direction will be equal to the velocity k with which the water-level, observer C, is rising. And it will be the Minkowski unique, limiting velocity. The velocity with which the point in the slit on page 77 moves spatially is, evidently, the same k; and this we, employing B_1's scale, should discover to be the velocity of light.

Hence it suffices to have a 'now 2' (typified by the water surface) travelling in the T_2 direction, in order to produce, for observer B_1, those relativity effects which would appear to him to be *as if* the vector in T_1 had been rotated. The same result would appear if we began by considering a 'now 3' travelling in a T_3 dimension, or a 'now 50' travelling in a T_{50} dimension. But the effect appears in the simplest way when we consider merely a 'now 1' travelling along T_1.

CHAPTER XVI

THE PHYSICAL OUTLOOK OF OBSERVER 2

The quantities which are considered in the problems of classical dynamics are :

Space, indicated by S,*
Time, indicated by T,
Mass, indicated by M,
Force, indicated by P.

It is convenient, sometimes, to represent space traversed per interval of time, or S/T, by V meaning velocity.

The way in which these quantities are interrelated is indicated in the following equation:

$$P = MS/T^2 \quad \ldots\ldots(4).$$

If we multiply both sides of this equation by T, we get

$$PT = MS/T = MV \quad \ldots\ldots(5)$$

This quantity MV, equal to PT, specifies the dimensions of the 'Momentum' generated in the moving mass in the course of the time during which the force acts upon that mass.

Instead, however, of multiplying the two sides of equation (4) by T, we may choose to multiply them by S, which gives us

---

* The more common practice is to denote space by L, meaning length, but to change to this symbol now might confuse the lay reader.

$$PS = MS^2/T^2 = MV^2 \qquad \ldots\ldots\ldots\ldots(6)$$

This quantity, MV^2, equal to PS, specifies the dimensions of the *'Energy'* generated in the moving mass in the time during which the force acts.*

Finally, let us multiply both sides of equation (4) by ST. The result is

$$PST = MS^2/T \qquad \ldots\ldots\ldots\ldots(7)$$

This quantity, PST or MS^2/T, is called *'Action'*, and it is a quantity of unique interest. A long time ago, it was discovered that all the laws which govern the paths by which a system changes from one configuration to another could be regarded as mere derivatives of a single general law that the action involved in such a change must be the least possible in the circumstances. This 'Principle of Least Action' was said to govern everything in physics from the path of a planet to the path of a pulse of light.

Clearly, we can regard this curious quantity PST as $PT \times S$, that is to say, as momentum multiplied by space. Or we may regard it as $PS \times T$, which is energy multiplied by time. This last way of regarding the quantity in question brings to light very clearly the most interesting feature of action. For energy, PS, is three-dimensional; and, when this is multiplied by T, the result is four-dimensional. Thus, action is a feature of a four-dimensional world, a feature which a three-dimensional observer divides up into components of energy and time.

Glancing through the foregoing equations, the reader will note that they exhibit the interrelations of what are, really, two systems of units. We can express all our problems in terms of the three dimensions P, S and T, or, equally well, in terms of the three dimensions M, S and T. Equation (4), viz.,

$$P = MS/T^2$$

which may be written also

$$M = PT^2/S \qquad \ldots\ldots\ldots\ldots(8),$$

provides the connecting link between the two systems. The first form of this expresses P in terms of the MST system, i.e., represents force as a name for mass × acceleration (S/T^2 is acceleration). The second expresses M in terms of the PST system.

The MST system has the illusory advantage that M, meaning 'mass', may be confused with the philosopher's 'matter' located at a definite place in space. Actually, the equations tell you nothing about the position of the 'matter'—unless you have agreed, previously, to accept the idea that the mass of

---

* Numerically, mv^2 is the *'Vis Viva'*, or twice the energy; but it is, consequently, proportional to the energy, and the numerical factor is of no importance in the present calculations.

a 'piece of matter' is located at the centre of gravity of the latter. Apart from presuppositions of that kind, neither system, in pure dynamics, makes any reference to matter. In the *MST* system the *M* is situated at a marked point in space : in the *PST* system the *P* is applied to a marked point in space.

The *PST* system, however, has a real advantage of simplicity, as the following table will show.

	PST system	*MST* system
Momentum	PT	MS/T
Energy	PS	MS^2/T^2
Action	PST	MS^2/T

Consider now the case of a classical physicist who is watching the behaviour of his B_1 instrument and is inferring from this the character of the ultimate object world. He would map out that world as a four-dimensional structure with time as a merely imagined fourth dimension (such as one sees in a barometric chart). But he would be quite unaware of the necessity of regarding his B_1 instrument as travelling along that dimension over an object system extended therein. Consequently, to him, time would be a simple scalar quantity; and it would appear as this in all his physical expressions. But, in *our* four-dimensional continuum, some of the physical quantities are different. Those which consist of P and S or combinations of P and S, where S is any one of the three dimensions of 'ordinary' space, change the S into $S/\sqrt{-1}$ which we may abbreviate into S/i While, wherever a classical physicist would write T, we should write cT_2. Making these alterations wherever necessary, we find that what, in the external world,

	The classical physicist regards as observed	We regard as measured
Time	T	$cT_2 = S_4$
Velocity	$V = S/T$	$V_2/ic = S/iS_4$
Mass	M	$ic^2 M_2$
Momentum	PT	$cPT_2 = PS_4$
Action	PST	$cPST_2/i = PSS_4/i$

where T_2, V_2, M_2, P and S are observer 2's names for quantities observed by B_1.

In what we may call the 'original' theory of Relativity, it was pointed

---

* The *i* attached to this *S* scalar quantity is not a "versor" but a numerical element in *S's* length.

out by Einstein that mass in the four-dimensional world must be mass multiplied by c^2. No satisfactory explanation was given as to why, in that case, this Mc^2 is observed by our instruments as plain M—it was inferred that Mc^2 must be energy relating to some 'internal' turbulence or what-not of the atom—an internal energy which no instrument could observe. But the important—and self-contradictory—inference which was drawn was that mass and energy were one and the same thing. We can tell a story more rational than that. Energy $i M_2 c^2$ is merely energy along the fourth dimension due to the relative velocity c existing between the instrument and the substratum.

The reader may wish to know, here, whether this relative velocity can alter. The answer is that the formula for lengths in the four-dimensional scale, $s_4 = ct_2$ makes those lengths dependent upon c. If c becomes less, the distances which we mark on the S_4 scale whenever our travelling clock ticks would become shorter, while our space units chosen earlier would remain unaltered. Consequently, the inclined world-lines (the positions of which are independent of how B_1 regards them) would indicate to B_1 that distances as before were being traversed in three-dimensional space, but that, now, a larger number of seconds were being taken over the journey. Thus, the effect of reducing the velocity of the 'now' would be to reduce all the velocities observed by the instrument, including that limiting velocity which is always the velocity pertaining to light. Hence, you can see whether the velocity of your 'now' is slowing down by seeing whether the velocity of light is diminishing.

It is quite evident from observation that this velocity does not vary every time B_1 transfers energy to, or receives energy from, A_1.

Now we, acting first as C, can begin to fill in our table.

A_1 is, to B_1, the content of a field of three-dimensional observation. This content appears to B_1 as changing; but the history of those changes is to be mapped out (says B_1) in a time dimension, and the field contains merely an instantaneous view of its content. Using the *PST* system, B_1 would call A_1, *PS*—the resistance encountered by a dynamometer multiplied by the distance S that the point of application has moved since the last observation. But C knows that B_1's supposed *PS* is really *P* x *iS*.

Next, we have to fill in A_2. That is easy : A_2 is the temporal history of the changes of that real *PS* from which B_1 abstracts *iPS*. The Victorian physicist would have written it *PST*. The man who, while admitting the travel of the instrument along time 1, fails to realise that the resulting map involves a right-angled revolution of a vector in T_1—this man, while describing A_2 wrongly, would describe A_2 nevertheless as being *PS* x *cT* = *PS* x S_4. With this we, acting for C, agree.

B_1 (says C) has the dimensions of a cross-section of A_2, namely, *PS*. It is only because he is travelling that he abstracts *iPS*.

I show, for purposes of comparison, our table and the table in which the revolution of the time vector has been overlooked.

Vector rotation overlooked		Our table	
A_1 PS		A_1 iPS	
A_2 PS x S_4	B_1 PS	A_2 PS x S_4	B_1 PS

It will be noticed that the only difference between the two tables lies in the fact that we write iS in place of S in our A_1.

If, now, we add to our table the proper description of C, we shall have carried the regress far enough. Repetition of the relations already discovered will have commenced. But D is now the describer, and he regards C as travelling. D's table would run

A_1 iPS		
A_2 PS x iS_4	B_1 iPS	
A_3	B_2	C PS x S_4

If we wish to fill in A_3, that task presents no difficulty. A_3 becomes

A_3 PS x S_4 x S_5 (infinite)

Note that it is now C, the only element that D regards as really travelling, who is made responsible for supposing the iS in A_1 to be S. He makes the same error with regard to A_2 and B_1 *(vide* his table) —an error which D corrects.

There is, of course, no A_3 for the man who has neglected the regress of time. His world is confined to A_2 and C, with light behaving quite madly in the former. The absence of a B_1 (whom he would have to regard as travelling) causes the most hopeless confusion, and will render his case quite desperate when he is confronted with modern 'quantum' physics. In fact he has made a thorough mix-up of his universe, and no amount of mathematical subterfuge will serve to conceal the wreckage.

Returning to the smoother pathways of the serialist, we can fill in B_2.

$$B_2$$
$$PS \times \sqrt{(S_5^2 + S_4^2)}$$
$$\text{finite} = \text{an action}$$

I have carried the regress far enough to bring in B_2, because that diagonal line in FIGURE 11 serves a useful purpose later on in affording a graphic explanation of that discreteness of Action, which is the real puzzle in what is known as the 'Quantum of Action'. The remainder of that puzzle seems to be merely part of the general mystery of why discrete things in Nature are atomic.

Where does mass enter into all this? Why, we can always substitute MV^2 for PS. For M in our table, the M_1 of the external world, becomes iM_2c^2, and V_1 becomes V_2/ic; so, naturally,

$$iM_2c^2 \times V_2^2 / i^2c^2 = M_2V_2^2 / i = PS / i$$

But mass—plain mass without adjectival trimmings—is not observed by B, It is an inference, and a very elaborate one, from observation of MV^2. This we shall see when we come to deal with the physiological aspects of the regress.

It is clear that this physical regress will proceed on the lines sketched out for as far as we care to carry it. But there is nothing to be gained by analysing it beyond the second-term observer C. The remainder will be mere repetitions exhibiting that relation between observer, self and object world which has been exemplified already in the table which contains C.

* * * * * *

We have finished with relativity for the moment. Our serialism has shown us why it is that $\sqrt{-1}$ is bound to enter into all relativity (and, for that matter, all atomic) calculations. Briefly, we cannot get Minkowski's world except by rotating the vector in a dimension of time 1 so that this vector coincides with an axis of three-dimensional space. When that is done, the picture in four dimensions appears as one which has been mapped out from observation of a three-dimensional instrument system which is travelling over the fourth-dimensional extension of the object world, and it becomes obvious that the *velocity—whatever its value—of* that travelling will produce, in the three-dimensional world apparent to the instrument, an equal and limiting velocity with all the remarkable attributes of the velocity of light.

CHAPTER XVII

QUANTA, WAVES, PARTICLES AND THE UNCERTAINTY PRINCIPLE

On December 14th, 1900, Dr Max Planck of Berlin announced to the German Physical Society his discovery of a strange new constant which he symbolised by the letter h. It became apparent very quickly that this h was nothing less than an *atom of action*—an atom of *PST*. It is known now universally as Planck's *'Quantum'*.

Planck had been studying radiation, and what his experiments proved may be explained quite simply. If we make the T in *PST* represent the period of the oscillation of a wave, we can say, obviously, that

$$PS = PST/\text{Period},$$

which means   Energy = Action / Period .

Planck showed that the action on the right-hand side of the equation must consist of indivisible atoms. Since fractions of these atoms could not exist, the equation must take the form

$$\text{Energy} = nh/\text{Period} \quad \ldots\ldots(9),$$

where n *is* some whole number and h *is* the atom of action—the quantum.

If, instead of regarding T as period, we regard S as wave-length, it is clear that

$$PT = PST/\text{Wave-length}$$

or   Momentum = Action / Wave-length

which Planck's discovery compels us to write

$$\text{Momentum} = nh/\text{Wave-length} \quad \ldots\ldots(10).$$

It is to be noted that these two equations (9) and (10) do not allow us to regard either energy *PS* or momentum *PT* as atomic. For period in (9), and wave-length in (10) are both variables. But the non-atomic quantity of energy which is equal to one atom h divided by the period of oscillation has proved to possess an importance equal to that of any atom. It is called, nowadays; a *'photon'*; and it is a well-established law that, in all interaction between an observing instrument and the object observed, what passes is energy to the extent of one or more photons. Moreover, Einstein showed, early in the century, that each 'photon', h/period , must arrive at the receiving instrument in the form of a particle travelling like a bullet, and not in the form of a wave. Nothing which did not possess these bullet-like characteristics could produce what is known as the 'photo-electric' effect.[*]

But (the reader well may ask) if these photons are *particles* possessed of varying amounts of energy, what is the meaning of 'period' in the

---

[*] The reader will find a very clear and simple account of this effect in the last chapter of Sir William Bragg's *The Universe of Light*.

definition of a photon as h / period ?

Well, the trouble was that, if you exposed a photographic plate to a direct beam of light, nothing but particles would arrive; but, if you passed that beam first through what is known as a 'grating', the effect produced would be exactly the same as if that beam had consisted of nothing but spreading light-waves. These waves would have length and period. The photons, on the other hand, had energy and momentum. And the law which emerged connected the light-particle in the one experiment with the light-wave in the other by the two equations (9) and (10) amplified thus:

Energy of the light-particle = h / Period of the light-wave
Momentum of the light-particle = h / Length of the light wave

Now, Newton had held that light consisted of particles shot out from the source in all directions. His contemporary, Huygens, proposed a 'pulse' theory, which, when modified and extended by Young and Fresnel, became the wave theory. This, in the interval before the arrival of Planck, held the field. The crucial experiment was the 'diffraction' of light by means of the 'grating' mentioned above. A 'grating' may be thought of most simply as an obstacle which hinders the passage of the light except through little apertures left open for the purpose. When a wave is checked by such an obstacle, any portion of it which arrives at a hole passes through that hole intact, but thereafter spreads out as a semi-circular wavelet radiating from the hole as a centre. Spreading thus from all the apertures in the grating, the wavelets cross one another's paths. Now, when two waves cross, and the crest of the one happens to coincide with the trough of the other, the result is to cancel the wave motion completely. If, however, the crest of one happens to coincide with the crest of the other, the wave effect is increased. The wavelets radiating from the holes fall on all parts of the receiving screen, but the part which is nearest the wavelet starting from one side of the grating is farthest from the wavelet starting from the other side. Thus the screen is struck in some places by wavelets which are in step, in others by wavelets which are completely out of step, thus cancelling one another, and in other places by wavelets which are partly out of step. The result is to make upon the screen a curious pattern of alternate light and darkness—the 'diffraction' pattern. It seemed incredible that any shower of particles could produce such an effect,—and the wave theory won the day.

The discovery of the photo-electric effect equalised matters. If particles could not account for diffraction patterns, waves could not produce the result we perceive after we have pressed the button of our Kodak.

The fact that light-particles could behave as waves suggested, of course, that particles of all kinds might possess this curious character; and, in due course, a wave theory of matter in general came into being. It was produced first by de Broglie, and presented later in an improved form by Schrödinger (who had arrived at it quite independently). Dirac may be said to have completed the work.

In the wave theories, particles are merely wave groups, analogous to patches of rough water in a sea. The waves of which these groups are composed may extend, theoretically, throughout the whole of space; but they neutralise each other everywhere except just in the region of the stormy patch. Such a wave group will, in most cases, travel at a velocity different from that of the actual waves of which it is composed.

It is almost impossible to analyse into distinctive classes the philosophical attitudes adopted by physicists towards these 'waves'. But one can trace a hazy division between two main schools of thought.

The first school regarded the waves as real, and the 'particle' as being merely a name for the wave group. Waves looked at from this point of view might be called 'metaphysical waves'.

The second regarded the particle as the underlying reality, and the waves as purely epistemological, i.e., as mathematical illustrations of the observer's ignorance concerning the present position of the particle.

The objection to the first attitude was insuperable. Nothing could prevent these wave groups from expanding. The expansion might be slow; but, even at its slowest possible rate, it would be too fast to permit of the existence of the world as we find it to-day. To quote C. G. Darwin: 'Even if we regarded the world as originally created in well-defined "wave-packets", they would certainly by now have spread indefinitely. We may say that the existence of fossils which have preserved their form unchanged for several hundred million years disproves the adequacy of the wave theory'.

The epistemological wave, or, as it was called, the 'probability wave-packet', was free from this objection. If the particle was travelling at an unknown speed in an unknown direction, our ignorance as to its whereabouts would increase with increasing time, and the area which might contain it would increase as the area of a packet of real waves would increase. Furthermore, the chances of finding the particle at any point in that area would be exactly equal to the *'intensity'* of an imagined expanding wave-packet at that point. An experiment which discovered the true position of the particle would bring the uncertainty to an end, and the wave-packet of purely imagined waves would be reduced suddenly to the tiny area occupied by the real particle. The objection that the troughs of the waves would have to represent 'negative' probabilities was an awkward one, but it seemed less overwhelming than the objection to the notion that the wave group was real, and yet shrank suddenly every time an experiment was made to ascertain whether it was, in fact, a particle.

These questions became acute when it was found that, just as in the case of the alleged light-particles, electrons could produce a 'diffraction' pattern.

I do not propose to drag the reader through the technical details of the various experiments which exhibited the apparently dual character of any alleged particle. He will find most excellent and lucid descriptions, abundantly illustrated, in Sir William Bragg's *The Universe of Light;* while C. G. Darwin's invaluable book, *The New Conceptions of Matter,* will show him precisely how the two classes of experiment—those which discover particles, and those which exhibit waves—are interrelated. One can summarise the empirical evidence as follows.

(1) Alleged particles shot against a screen coated with zinc sulphide crystals will produce tiny sparks at the points where they strike the screen, showing, thus, the strictly localised character of the collision.

Alleged particles shot through a Wilson cloud chamber cause condensations of moisture along the tracks of the supposed tiny bodies. These tracks indicate that what has passed is something very small which is travelling in space in a perfectly normal fashion.

(2) Showers of alleged particles falling on a photographic plate after they have been interfered with by a 'grating' produce a diffraction pattern such as would be made by alleged waves.

(3) The two classes of experiment cannot be combined. It is impossible to discover, at one and the same time, both the 'particle aspect' and the 'wave aspect' of whatever may be the ultimate reality. Consequently, we cannot fall back upon the notion of a group of real waves containing a real particle.

The whole thing boils down to this: Set a trap to catch particles, and you will catch particles; set a trap to catch waves, and you will catch waves. And *all* the experiments appear to be crucial, ruling out definitely either one aspect or the other. This, to a serialist, gives rise to the suspicion that it may be the *nature of the experiment and not the nature of the object* which is really in question.

To the general cauldron of trouble we may add a couple of ingredients. The Schrödinger waves are not waves in space alone, but waves in space and time. Each electron requires the whole of ordinary three-dimensional space for its waves, and will not permit the presence of any other electron in that space. Two electrons require a space of six dimensions,—three apiece,—and so on. Which makes the serialist, with his mild regress of time dimensions, appear quite timid.

The reader must bear in mind the way in which the quantum—the atom of action—is involved in all these difficulties. The whole of the wave theory is dotted with h's. And h appears again in what is known as 'Heisenberg's Uncertainty Principle'— a principle which we must proceed now to consider.

Every experiment (as I have pointed out *ad nauseam*) *is* an interference with the object system by something three-dimensional which is regarded as separated from that system. Again, every collisional observation by a three-dimensional instrument involves an interchange of energy between the instrument and its object and is, consequently, an interference with that object. Now, Heisenberg remarked that what must pass between observer and observed in such cases cannot be less than, and cannot be dimensionally different from, one photon, h/period—which is the energy content of one atom of action h. Consequently, every measurement of action *PST* must lack precision to the extent of the amount contained in h.

Such a measurement would be, for example, a simultaneous measurement of *PT* and *S* in the case of a particle. The total uncertainty h in the amount of action must appear in the measurements of *PT* and *S* in such fashion that our uncertainty about the momentum of the particle multiplied by our uncertainty about its position cannot be less than h. In these calculations we write p for momentum and q for the coordinate giving the position of the particle at the moment of experiment. 'Uncertainty' is symbolised by Δ. So that Heisenberg's equation runs

$$\Delta p \, \Delta q \approx h.$$

($\approx$ means, 'is of the order of magnitude of'.)

This Uncertainty Principle appears to be absolutely inviolable, so we had better ascertain exactly what it means. Fortunately, the meaning is extremely clear and precise.

The impact of the apparatus for measuring velocity alters the velocity of the supposed particle to an unpredictable extent. The two measurements of position and momentum are supposed to be made simultaneously. Very well:

At that instant, the *present* position of, and the *past* velocity of, the particle may be determined with any degree of accuracy we please. The Uncertainty Principle does not apply to these two determinations. But...

At that instant, the more accurately we measure the *present* position of the particle the greater becomes the uncertainty in our knowledge of its *future* velocity, so that

$$\Delta \text{ present } q \times \Delta \text{ future } p \approx h.$$

All physicists, including Heisenberg himself, are agreed upon these two facts.

Now *whose is* the uncertainty? It will not be disputed that the observer is uncertain, so we can take that for granted and go on to the next question. Is there, *in this Uncertainty Principle alone,* the slightest shadow of an excuse for supposing that *there can be no such thing in the universe as a* particle possessing simultaneously both definite position and definite velocity?

I have tried to put that question plainly, but those who suppose that there are grounds for an affirmative answer are less explicit. 'It is the velocity after the measurement which alone is of importance to the physicist', says Heisenberg. Why? Is it not part of the physicist's task to explain what has happened—to show how such-and-such a situation has come about? Sir Arthur Eddington, again, remarks that the velocity which we ascertain by two successive measurements 'is a purely retrospective velocity'. But does that mean that our acquired knowledge thereof is to be ignored? If so, why?

The truth is that Heisenberg's Uncertainty Principle gives a plain answer to the question as to whether the Schrödinger 'waves' are to be regarded as epistemological or metaphysical. And the answer is against the metaphysicians.

For, suppose that the waves were objectively real. Suppose that Nature knew nothing of such things as particles. Then we should find that our supposed 'particle' was a figment of our gross imaginations, trained to the appreciation of a macroscopic (large scale) world. And, if we were foolish enough to insist that the wave-group exhibited nothing beyond our own ignorance of what we had done to the particle in the course of an experiment, Nature would give us the lie.

But her verdict would be retrospective.

There is no getting round that. In such circumstances, we should find that the alleged particle *had never possessed, at any time*, the two mutually exclusive attributes of precision in position coupled with precision in velocity. The wave-group would not have permitted it. We should find that the precision in velocity had always varied inversely as the precision in position.

Very well. I make six successive determinations of the position of a supposed particle; which determinations, according to the Uncertainty Principle, may be, theoretically, as accurate as I please. Each of these determinations, after the first, informs me of the velocity of the particle since the previous measurement was made. Each determination disturbs the velocity previously ascertained, but in each case, except the last, I am able to say exactly what was the extent and direction of that change in velocity. I have, therefore, a history of the particle showing that it possessed definite position and definite velocity on four occasions according to my opponents, and on five occasions according to myself. The admitted four occasions are sufficient for my purpose. Nature knew nothing then of *an Uncertainty Principle!*

She has heard of it since,—from the New Metaphysicians,—but is entirely unable to alter her distressing past. The most that she can do is to agree quickly that the metaphysician's knowledge as to what has become of the particle since the last time he hit it is mathematically representable by the intensity of a wave. She hopes profoundly that he will be satisfied with this makeshift and will probe no deeper into the matter.

He never does.

Conclusive ! Of course. But *all* the arguments in this *imbroglio* are conclusive. If it were not so, there would be no confusion. Here is a reply to myself. If the waves are merely imagined, how can they make a mark upon a photographic plate?

Note, please, that this is an instance of the way in which the dispute is carried on. No side can refute the arguments of its opponents—it has to content itself with advancing another argument of a totally different kind. In a copy of *Nature* which lies open before me, I find Sir James Jeans's announcement, to the British Association, of a supposedly crucial experiment which favours the wave; while Professor Andrade, on another page, is pointing out how the discovery of 'The New Elementary Particles' furnishes a final answer to the vexed question, and a verdict for the particle. But the experiments in the two cases were entirely different. And, until we understand a little more of what we are doing, we have no right to say that, in any experiment, the particle-picture and the wave-picture have 'come into conflict'. In other words, we have no right yet to presuppose that the trap which has caught a wave was a trap for particles, or *vice versa*—*nor* shall we have that right until we have made the trap the object of our observation.

That we shall do in the next chapter.

CHAPTER XVIII

THE REGRESS OF UNCERTAINTY

It will have been obvious to the reader that, in their interpretations of the Uncertainty Principle, the several parties concerned have been regarding the 'now' as all-important, and have been treating that 'now' as travelling in the fourth-dimension. Consequently, they are drafting their pictures in terms of an infinite regress. But to draw a picture of a certain kind while pretending to yourself that you are drawing something else is not the way to do full justice to your capacity as an artist. It is not surprising, therefore, that the picture has gone wrong.

This is what has been drawn. The artist starts with the state of affairs where a determination of the position of the particle is made. Then, whether he regards the particle as being really a wave-group, or believes the wave-group to be a mere abstract 'probability-packet', he marks out the future in time 1 as an area enclosed between two world-lines showing the limits of the changes which may have been made in the particle's velocity, and these lines show the way in which the wave-group expands in three-dimensional space. (For simplicity in the diagrams, we shall show these world-lines as extending evenly on either side of the time direction, *vide* FIGURE 19.)

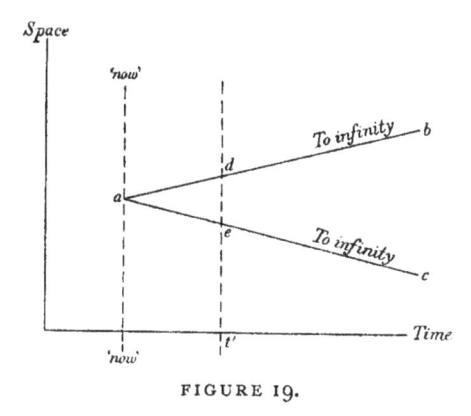

FIGURE 19.

Here *a* is (let us say) an electron. Its position is being determined within a small space area (represented by the thicknesses of the lines *ab* and *ac*). This determination disturbs its velocity. The artist's ignorance of the extent of that disturbance is of such a magnitude that, when he makes the next observation (at, say, any instant t') he may rediscover the electron anywhere upon the line *de*.

He proceeds then to picture this second determination of position as being made. That is to say, he considers the case where the 'now', *and, of course, his instrument,* (though he does not mention this), has shifted to t'. He supposes that the electron is rediscovered at, say, a point *f*, and he exhibits, in FIGURE 20, the resulting situation.

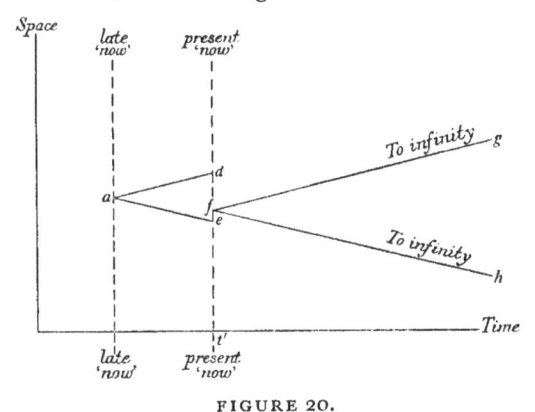

FIGURE 20.

At this stage, the notion that the wave-packet is real begins to look absurd. For the new disturbance given to the rediscovered electron could

91

not cause an expanding group of real waves to contract instantaneously to a tiny area in the manner shown.

How do the advocates of wave reality get over this difficulty? I cannot tell you. At this juncture they cease to talk about waves, and commence a dissertation upon the inadequacy of space and time descriptions and the folly evinced by man in supposing that Nature would allow herself to be described in terms suitable to his gross mind—this last being a theme in which they feel really at home.

That plea, as always in the history of mankind, proves to be inadmissible. We are crying out before we are hurt.

The idea is that the real-wave theory proves adequate up to a certain point, and then breaks down. Also, that the particle theory proves workable for a little while, and then collapses. But, in the picture we have shown, the particle theory does not fail anywhere—if the wave-packets are only areas exhibiting the ignorance of the experimenter at the 'now', an ignorance which, subsequently, is enlightened. There is no collapse of the particle-picture so long as you content yourself with seeking for the position of the particle. It is not until you introduce an experiment which seeks for waves that the trouble begins.

Now, it will be obvious to any serialist that FIGURE 20, as an illustration of two successive happenings at the 'now', has been wrongly drawn. It requires the introduction of another time dimension in which to exhibit the changes in position of that 'now' and of the instrument of discovery which travels therewith. That we will deal with in good time. But I want to point out that the result is to obscure a fallacy in the picture of the past. For the experimenter is seeking for, and discovering, the particle, and is making no other kind of experiment. He has no reason, therefore, to exhibit his past wave-packets as having been *anything in the 'substratum'*—*anything* pertaining to the object observed. They were memoranda of his own ignorance, an ignorance which has been enlightened when the experiment at t' is made. The correct picture would have been as in FIGURE 21. It represents the kind of time 1 map of the electron's career which could be drafted from the information provided by a series of scintillation experiments or from observation of the track in a Wilson chamber. Only one past position of the electron is shown, but there is no reason theoretically why the past part of the picture should not show a whole series of past positions of the particle and the knowledge of its velocity obtained from these, precisely as I indicated in the imagined experiment of the last chapter.

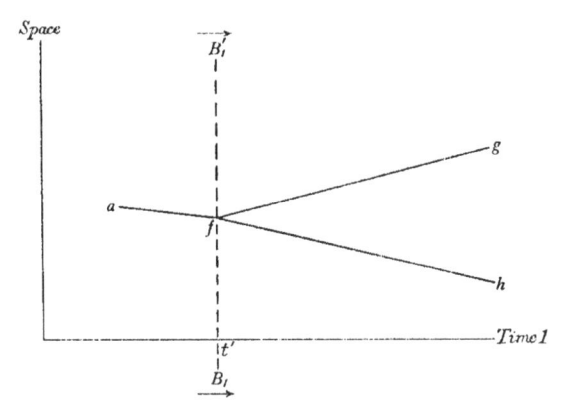

FIGURE 21.

Now, in the ordinary course of exhibiting a time regress, the next stage is to draw a diagram which shall include the instrument B_1 and map out the successive positions of this, employing another dimension for ultimate time and treating the T_1 axis of FIGURE 2I as an axis of S_t. But, before we can put the instrument into any such picture, we must note what the Uncertainty Principle has to say about that instrument.

Of Heisenberg's many illustrations, the one quoted most frequently is the famous imagined experiment with a microscope. The apparatus is supposed to be an adjunct to an eye observing an electron by means of light scattered from the latter. Heisenberg considers the cone of rays scattered from the electron and entering the aperture of the instrument as yielding the necessary information about position q. He then considers the recoil which the electron receives from this light; and, for that purpose, assumes that one photon of light passes. He relates the momentum of this photon to the wave-length of the light-waves entering the aperture by the formula (see equation (10)) $p = h/\lambda$, where p is momentum and λ *is* wavelength. He has no difficulty in showing that the uncertainty in the determination of present position is related to the uncertainty of the future momentum by the equation $\quad \Delta p \, \Delta q \approx h$

The example is not a very good one, and I quote it merely because of Heisenberg's concluding remarks, which I give in full below.[*]

'Objections may be raised to this consideration; the indeterminateness of the recoil is due to the uncertain path of the light quantum' (i.e., photon) 'within the bundle of rays, and we might seek to determine the path by making the microscope movable and measuring the recoil it receives from the light quantum. But this does not circumvent the

---

[*] *The Physical Principles of the Quantum Theory,* by Werner Heisenberg. (Cambridge University Press.)

uncertainty relation, for it immediately raises the question of the position of the microscope, and its position and momentum will also be found to be subject to the equation

$$\Delta p \, \Delta q \approx h$$

The point to be noticed in this imagined extension of the experiment is that *when we put the instrument into the picture, as B_1, and observe this from the viewpoint of C, we transfer the uncertainty of p and q from the original object electron A_1 to the instrument B_1. We exhibit our uncertainty regarding A_1 as being due entirely to our uncertainty concerning B_1, and not to anything intrinsic in the character of A_1. We are not confronted then with both an indeterminate electron and an indeterminate instrument, which would give more uncertainty than the quantum restriction h permits.*

It will be perceived that, in this imagined extension, the microscope is supposed to be actually *recording* the momentum received from the electron (strictly speaking, of course, from the photon). The C which observes the instrument's observations of the electron (records both the light coming from the eyepiece and the imagined motion of the eyepiece) could be, e.g., a strip of sensitised film. But the illustration, as said before, is not a very good one: the experiment is impracticable; and the change in the momentum of the microscope would be inappreciable, owing to the large mass of that instrument. We will pass on, therefore, to Heisenberg's analysis of a real experiment, viz., the scintillation produced by the impact of an alpha particle upon the surface of a prepared screen.

The scintillation is produced by the 'ionisation' of an atom in the prepared screen, that is to say, the incident particle knocks an electron in the screen out of the orbit in which it is circulating within the atom. That orbit constitutes a slightly hazy point in our mesh system, (the screen), hazy because we do not know the position of the target electron within that orbit. The momentum of the incident particle is changed, of course, by the impact.

How are we to measure that change in the alpha particle's momentum? Clearly, whatever momentum it loses is transferred to the electron ejected from the atom. Now, we can measure the momentum of the ejected electron precisely, after it is ejected. But the trouble is that we do not know what was its momentum before it was struck. Thus the uncertainty in the position of the incident alpha particle is due to the uncertainty of the position of the *instrument* electron within its orbit; and the uncertainty in the new momentum of the alpha particle after the collision is due to the uncertainty of the momentum of the *instrument* electron within that same orbit. Heisenberg, explaining this in slightly more condensed language, and taking the nature of Bohr orbits into consideration, relates these two uncertainties by the equation

$$\Delta p_s \, \Delta q_s \text{ is not less than } h,$$
where the little *s* refers to the orbit of the *instrument* electron.

But when the two uncertainties are regarded thus as pertaining to the instrument, the alpha particle is being assumed to possess a perfectly definite track both before and after the collision; that is to say, there is not supposed to be any *intrinsic* uncertainty in its behaviour. To assume the contrary, while allowing for the two uncertainties in the instrument, would give us more uncertainty than *h* can provide.

So, in this experiment, again, *putting the instrument into the picture*, as a B_1 *observed by a C*, transfers the uncertainty from A_1 to B_1.

Similar considerations apply, of course, to the ionisation of an atom in the Wilson cloud chamber experiments.

Now we know where we stand, and we can get on with a description of the kind of time map which would be drawn by our imagined C.

He is a four-dimensional observer with a field of observation extending the whole length of A_2, which constitutes his 'now' in a world where time is a fifth dimension, T_2. B_1 is an object at the point O, and has just been employed by C as an instrument for obtaining information about the substratum at that point, i.e., information about A_1. We saw earlier (p. 64) that C, being a four-dimensional observer, cannot distinguish three-dimensional sections of the substratum without the assistance of B_1. (B_1, since it is travelling at the velocity of light, c, has, to C, no fourth-dimensional extension.)

In C's world, consisting of A_2 and B_1, there is no inherent uncertainty. The particle disturbed by B_1 has a perfectly definite world-line both to the left of and to the right of (in our maps) the point of impact. The trend of the line to the right of that point, i.e., in the time 1 'future', is altered by that impact—altered instantaneously in fifth-dimensional time. Let us suppose that this disturbance of the particle at O has repercussions in the objective world,—produces, for example, an explosion,—and alters, consequently, the general character of that substratum to the right of O. That change would be apparent to our imagined four-dimensional observer C. And his A_2 world, which is Nature's world, would be recognised by him as perfectly 'determinate' so far as the pseudo-time, time 1, is concerned. To C, the fifth dimension T_2 is time, and the four-dimensional world is, simply, 'present', and equally definite everywhere. But B_1's future does not lie in that A_1 world. B_2 *is* a world-line (the $O'O''$ of FIGURE 11) which intersects A_2 at only one point. So that B_1's future lies outside C's view. Now, we have just seen that, according to Heisenberg, putting the instrument into the picture as something observed transfers the uncertainty from the original object particle to the particle in the instrument. C, then, is uncertain as to the future of B_1. He does not know precisely what has been the change in its velocity in three-dimensional

space (the space in which the impact occurred). He cannot map out the trend of its world-line along the four-dimensional stretch B_2. And his uncertainty is governed by the rule
$$\Delta p \, \Delta q \approx h$$
just as in the case of B_1's uncertainty about the future of the original object particle in A_1.

Note that in both cases the uncertainty is the same. It is an uncertainty as to whereabouts in ordinary space *the instrument will encounter the particle* in a future experiment. But the correct development of the regress shows this, first as an uncertainty regarding the future position of the particle as referred to B_1, and then, in the all-important second term, as an uncertainty in the future position of B_1 as referred to C—the time 1 future of A_1 being certain as referred to the C system.

It is clear that if we put C into the picture we shall find that the uncertainty of our knowledge concerning B_1 is due entirely to the uncertainty of our knowledge concerning C. The observer who puts C into the picture is D. The map he would draw of space and time (time being the sixth dimension to him) would show both B_2 and A_3 as having definite position in the 'present' five-dimensional world, but it would show the future of C, which is in the sixth dimension, as having the quantum uncertainty.

Thus, the uncertainty recedes up the ladder of the infinite regress. It is an uncertainty about the unreachable absolute future. But, in the second term and onward, we discover that it is *an uncertainty pertaining only to the last instrument in the picture and never to the world which we are studying by means of that instrument.*

What alterations do we require to make now in FIGURE 11?

As presented, the diagram is a picture of the world *observed* by observer 3—D in the table. It is he who observes C as a travelling instrument, and his uncertainty is an uncertainty about the future positions of C in three-dimensional space. *That* future is not in the diagram.

If, however, we wish to treat FIGURE 11 as a picture drawn by C, including the future as *calculated* by C from his knowledge of the present world A_2 (or GH), we should need to draw OO'' dotted, in order to indicate C's uncertainty about its future spatial position. But $O'O$ is a determinate line, and should be drawn as before.

CHAPTER XIX

THE WAVE EFFECTS

We have to reply now to two questions, viz.:
(1) Can we *prove* this regress of uncertainty—prove it by actual experiment?
(2) What about those wave effects?
The answer to the first question is, 'Yes': the reply to the second is that it is the wave effects which constitute the experimental proof required.

We are going to investigate the nature of light. A beam of light is, consequently, our A_1 object. For our B_1 instrument, we shall employ, instead of a scintillating screen, a complete diffraction apparatus comprising a ruled metallic reflecting grating, (this diffracts just as well as a transparent plate with opaque rulings), and a photographic plate to receive the rays after their reflection.

The result of the experiment will be the appearance of a diffraction pattern on the plate. Our business is to ascertain what must be the nature of the rays which produce that effect.

Our scintillation experiments have taught us that the beam of light consists of a shower of particles. Since those experiments were more direct and simple than the one on which we are engaged now, we shall begin by seeing what would happen to a shower of particles striking the grating and being scattered in all directions. On this particle theory the diffraction effect must be due entirely to that scattering; for light particles do not interfere with one another when their paths cross, because they carry no electric charge. So what we have to study is the nature of the interaction between the particles and the ruled reflecting surface.

Now, we know the position of the apparatus in our laboratory and can regard both laboratory and apparatus as a single spatial system. We know the width of the beam of light relative to that system. But we have not the remotest idea whereabouts in that beam is any individual particle. This is a considerable uncertainty in our knowledge of the position of the point where that particle strikes the grating. But position is relative, and we can express this uncertainty in two ways. We may say either that we do not know the position of any particle relative to the screen, or, equally well, that we do not know the position of the screen relative to any particle. We will interpret the uncertainty in the second of these two ways. It is very considerable: let us see if we can reduce it.

The demonstration which follows is Duane's, and is one of the prettiest bits of work in the whole of mathematical physics. But the nonmathematical reader, I fear, will be unable to follow it for more than a little way. Still, the general idea will be apparent to him, so he should skim through the text. For the rest, he will have to be satisfied with the fact that the demonstration is accepted, and quoted with approval, by Heisenberg, who adds interesting comments.

It turns out that we *can* reduce the uncertainty. For, suppose we were to move the grating. A movement of the whole grating to the extent of the distance between the rulings would not affect the diffraction; for a particle which, before the movement, would have fallen on one ruling, would fall, after the movement, on another ruling in the same place as the first, so that the diffraction effect would be unaltered. This critical distance between the rulings is called the grating 'constant'. We will symbolise it by *d*. The dimension in which such movement could take place, at right angles to the ruling, we will call *x*. I will continue now in Heisenberg's own words.*

'Translation in the *x*-direction may be looked upon as a periodic motion, in so far as only the interaction of the incident particles with the grating is considered; for the displacement of the whole grating by an amount *d* will not change this interaction. Thus we may conclude that the motion of the grating in this direction is quantized and that its momentum p_x may assume only values nh/d (as follows at once from the earlier form of the theory : $\int pdq = nh$).'

Note that this introduces the quantum as an atom of action but not yet as a connecting link between wave and particle. That is what has to be proved. Heisenberg continues :

'Since the total momentum of grating and particle must remain unchanged, the momentum of the particle can be changed only by an amount mh/d (*m* an integer) :

$$p_x' = p_x + mh/d$$

Furthermore, because of its large mass, the grating cannot take up any appreciable amount of energy, so that

$$p_x'^2 + p_y'^2 = p_x^2 + p_y^2 = p^2$$

If θ is the angle of incidence, θ' that of reflection, we have

$$\cos \theta = p_y/p \quad \text{and} \quad \cos \theta' = p_y'/p$$

whence

$$\sin \theta' - \sin \theta = mh/pd$$

The rest is simple. We can write the above equation in the form

---

The Physical Principles of the Quantum Theory, (Cambridge University Press.)

$$d(\sin\theta' - \sin\theta) = m \times h/p$$

But, in the ordinary wave theory,

$$d(\sin\theta' - \sin\theta) = m \times \lambda$$

therefore $\quad\quad\quad\quad h/p = \lambda$

That is to say, from an inspection of the pattern on the plate a length can be arrived at, really a measure of h divided by the momentum of the particle, which length would be equal to the wave-length of the particle had the grating been treated as of fixed position and had the particle been a veritable wave.

The following comments are, I believe, pure Heisenberg; but I apologise to Duane if I am mistaken.

'The dual characters of both matter and light gave rise to many difficulties before the physical principles involved were clearly comprehended, and the following paradox was often discussed. The forces between a part of the grating and the particle certainly diminish very rapidly with the distance between the two. The direction of reflection should therefore be determined only by those parts of the grating which are in the immediate neighborhood of the incident particle, but none the less it is found that the most widely separated portions of the grating are the important factors in determining the sharpness of the diffraction maxima. The source of this contradiction is the confusion of two different experiments. If no experiment is performed which would permit the determination of the position of the particle before its reflection, there is no contradiction with observation if the whole of the grating does act on it. If, on the other hand, an experiment is performed which determines that the particle will strike on a section of length Δx of the grating, it must render the knowledge of the particle's momentum essentially uncertain by an amount $\Delta p \approx h/\Delta x$. The direction of its reflection will therefore become correspondingly uncertain. The numerical value of this uncertainty in direction is precisely that which would be calculated from the resolving power of a grating of $\Delta x/d$ lines. If $\Delta x \leq d$ the interference maxima disappear entirely; not until this case is reached can the path of the particle properly be compared with that expected on the classical particle theory, for not until then can it be determined whether the particle will impinge on a ruling or on one of the plane parts of the surface, etc.'

We need not, in this experiment, trouble about the uncertainty of the positions of the individual atoms of the apparatus. We are dealing with an uncertainty so large (the whole width of the grating constant) that the atomic uncertainty is negligible.

Now, we have regarded the position of the instrument as uncertain by that large amount. The result is to produce a diffraction pattern, *provided that the light consists of perfectly determinate particles, behaving just as classical particles would behave.* For the momenta of the particles *before* impact are regarded as free from the restrictions of the h rule. That they arrive at the plate in a subservient condition, is due to their traffic with the atoms of action of the grating.

If, on the other hand, we regard the position of the grating as determinate, and not subservient to the h rule, we shall get the same diffraction pattern, *provided that the light particle is a merely imagined point in what is really a wave-group governed before impact by the quantum restrictions.*

The illustration is clear enough. Every uncertainty in Nature can be regarded as your own uncertainty concerning your instrument. The case here parallels on a larger scale the case of the scintillation experiments. There we saw that, if we assert that the uncertainty in the position and momentum of the ejected ionised electron follows the h rule, then the incident particle must be determinate and free from such restrictions.

The reader may be a little puzzled as to how we can contrive to construct a science if we have to regard our instrument as indeterminate. The answer is: Easily enough, if you know the rule governing that uncertainty—the h rule. He may wonder, also, whether it would not be simpler to treat the instrument as free from h restrictions, and to attribute these to the system under observation. But here the rule of the regress comes in. When any knowledge has to be expressed in the form of an infinite regress, you must trace that regress far enough to bring in the relation between the second term and the third. That means, in this case, that we must regard the universe from the point of view of a four-dimensional observer, who would put the instrument into his picture and regard that instrument as the only thing which is governed by the h rule. And remember: it is impossible to imagine a more effective way of losing knowledge, or a more prolific method of introducing confusion, than that which consists in expressing your knowledge in the form of an infinite regress and then confining your study to the first term alone.

If the reader has still any doubts remaining, let him glance at FIGURE 22. It exhibits the relations between the atom of action and the two uncertainties of position and momentum. I have copied it from a sketch I made last Spring; but the demonstration has been published independently since then by Professor Flint in the pages of *Nature,* where it elicited no contradiction that I noticed. So the reader may regard it as sound.

The entire area $ADGJ$ represents action pq. The small area N represents an atom of that action; and it will be seen that it is equal to $\Delta p \Delta q$. Thus, the uncertainty of an action measurement is due to the *atomicity* of the action. Obviously, if you regarded the uncertainty in the action measurement as due to the difference between the area of the whole figure *ADGJ* and the inner area *M*, a difference, that is to say, equal to the areas *ADFC+EFGH*, then the Uncertainty Principle would not hold. So far Professor Flint goes. What follows is my own opinion, but I do not anticipate any disagreement from so clear-sighted a physicist.

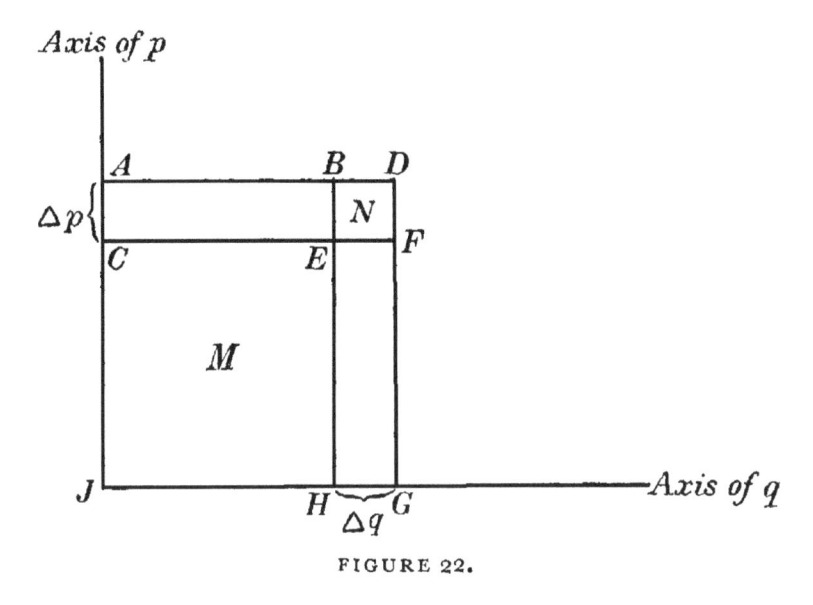

FIGURE 22.

Suppose we asserted that the instrument and the object measured thereby were *both* composed of atoms of action equal to the *N* of the figure. It is clear that the uncertainty in the resulting measurements of the object would be doubled. $\Delta p \Delta q$ would be $2N$. Can we get round this by supposing that N in each case $= h/2$, so that the sides of that area equal $\Delta p / \sqrt{2}$ and $\Delta q / \sqrt{2}$ respectively, instead of the Δp and Δq shown? No, for the total observed uncertainty in the measurements of p would be then

$\Delta p / \sqrt{2} + \Delta p / \sqrt{2} = \sqrt{2} \times \Delta p$

instead of Δp required by quantum theory—and so with Δq.

So we must have action atomicity either in the instrument or in the external world, but not in both. And, as already explained, the nature of the Time picture attributes that atomicity to the instrument. It is to be noted, of course, that, while *C* will regard B_1 as indeterminate, and A_1 (inferred as abstracted from A_2) as determinate, he will realise that the indeterminate character of B_1 will make B_1 observe A_1 as also indeterminate. Consequently, so long as observation is confined to a single A_1 only, and this is not interfered with—between observations—by other entities in the external world, no error will be perceived. But that would be a very limited kind of science.

The correct procedure for a modern physics which seeks to ascertain the nature of the external world is to assume quantum uncertainty in the instrument and *no quantum uncertainty* in Nature. Then, and then only, is it possible to calculate easily what is going on among the entities which are not being observed at that instant. That calculation having been made correctly, an experiment—in which, again, allowance is made for the instrument's uncertainty—will prove the accuracy of the work. When the instrument interferes, it passes an atom of action to the external world or accepts an atom therefrom, but there is no need for us to attempt the impossible picture of that atom maintaining its integrity in that external world. Indeed, the regress forbids us to entertain any such notion—forbids us to convert our epistemology into a metaphysics—forbids us to attribute to Nature an indeterminism which pertains, properly, to the observer.

The reader will appreciate now the significance of the warning given at the end of Chapter XVII.

CHAPTER XX

INTRODUCING THE REAL OBSERVER

The reader will appreciate now the complete artificiality of analysis in terms of time. I take two objects, both, to me, in the A_1 class, and—hey presto!—*one* of them turns into a B_1 galloping along the time in which the other one endures. It is purely a matter of interpretation, and the interpretation depends upon which one I choose to select as my source of information about the other. But the reader will have realised also, I hope, the extraordinary way in which this device abstracts sense out of what, otherwise, would be nonsense.

He will guess, moreover, whither the last paragraph is heading. I should like to hurry on towards that goal. But we cannot do that yet. There is a great host of objectors standing by—a host headed by the allied ghosts of John Locke and Ernst Mach—a host of innumerable epistemological purists.

Both Locke and Mach, I think, would have insisted that our journey has been made from a starting point which I omitted to define. For, at the beginning of Chapter VII, I opened the time regress in the following words:

'Let M represent a particular configuration of the external world as this last is described by you from observation, experiment and calculation. The particular configuration which M is to represent is the one which is open to your observation at the present moment.'

How are you to know *which is* this 'present' configuration? And what is the use of my telling you that you must put your chosen instrument at that 'now' in the time map, before you have discovered where that 'now' is? The instrument may *mark* it, when found; but, since you can change instrument and object about at will, neither of these can *make* it.

So the whole analysis has been based upon the presupposition that you, as a *psychological* individual, are situated at the 'now' of some time which is apparent to you. It has been founded, moreover, on the presupposition that you have knowledge of a physical world as well as knowledge of a world of phenomena. We must accept the first assumption, otherwise the whole physical demonstration breaks down. We must do something more than accept the second, if we are to construct an edifice which philosophers will regard as other than a phantasy.

Note that we have not got to justify the first hypothesis—your knowledge of a psychological 'now'. We are trying to discover whether there is any method of describing the universe which would satisfy the needs of the self-conscious observer we imagined in the previous chapter. We are proceeding by a method of trial and error. 'Here is time! Let us see if *that* fits.' So we try what amounts to equipping you with an intuitive knowledge of 'now'. The analysis in Part 11, 'General Test of the Theory', shows that this fits to perfection. It shows that anyone with the initial *intuitive* knowledge of a 'now' must have an *intuitive* knowledge of the serial dimensions of time, and can be a self-conscious observer.

Now, the original analysis of any self-conscious observer showed that such a creature would regard his *objective world* as comprehensible and as subject to his interference. So, in Part III, we tried equipping this psychological observer with an intuitive appreciation *of force,* as well as of space and of time. Possibly, you did not notice that we were doing this; but it was implicit in the statement that he could take P, S and T—instead of M, S and T—as *elementary indefinables* in terms of which the objective world could be described. It was proved thereafter that the world in question would be regarded as comprehensible. But the supposition of an intuitive knowledge of P, S and T as indefinables suited to the description of an external world of physics meant that, if the psychological observer possessed that intuitive knowledge, he could *discover* that physical world. This would be a reply to Subjective Idealism. Consequently, we must examine it rather carefully.

There are certain phenomenal objects, e.g., a chair', which, when you apply force to them, move. Given the intuitive appreciation of resistance and the intuitive appreciation of space, the resistance appreciated multiplied by the appreciated distance of displacement of the phenomenal object constitutes a complete appreciation of physical energy. The appreciation of this complex is not elementary,—it is a 'percept' and not a 'sensation',—but that is immaterial. External physical energy can be *discovered.*

Next, let us look at the matter from the point of view of psychophysiology. Among the various kinds of neurones with which your nerve endings are equipped, there are some which can be stimulated by simple *pressure*. These are to be found in the skin, in the muscles and embedded in those parts of the joints which roll upon each other. The pressure registered by the muscular neurones is a measure proportionate to the strain exerted by those muscles in moving a limb : the change in the pressure from one neurone to another in the rolling surfaces of the joint gives you direct information as to the amount of rotation of the limb. Consequently, when you move a limb, you can perceive *PS,* or energy.

In both cases the energy appreciated is a percept, and a percept which is just as much 'phenomenal' as is that percept of the coloured sphere which you learn to regard as an 'orange'. In both cases assimilation and association are at work to produce the complete percept.

Now, let us add the appreciation of time, *T*. Whenever you move a particular portion of your body, a curious law comes into operation; and this law is open to your appreciation. In all the changes of P, S and *T* accompanying the change of position of the limb there is one quantity which remains constant, and that quantity is the force divided by the acceleration. That quantity is the *mass* of the limb. The process of learning what force to apply in order to produce a required acceleration of the phenomenal limb (or acceleration of the rate of change of pressure from one neurone to another in the joint) is precisely the same thing as learning what is the mass of the limb involved. There is, then, no reason why a child in the pre-natal condition should not become aware of the world of mass.

And the, possibility of discoveries of this kind is not confined to the realm of the body. The pressure neurones in the skin of your finger tip will inform you of the resistance offered by an external object of which you have no other sensory appreciation. If you move the finger, the joint neurones inform you of the displacement of that point of resistance. But the pressure recorded will be less than the pressure recorded by the muscular neurones, because the pressure in the latter case is that needed to accelerate both the limb and the external mass, while the finger-tip pressure is that which is needed to produce the *same* acceleration in external mass only.

Thus, the intuitive knowledge of time and space accepted (on trial) in Part II, *plus* the sensation of pressure (demonstrable in any psychophysiological laboratory) provides any purely psychological observer with all that is necessary for the discovery of an objective physical world.

If the reader does not like this theory, he will have to fall back on one which is, I regret to say, rather popular nowadays. The idea is that the child distinguishes, after birth, phenomena appearing and disappearing at certain points in space; discovers, by consultation with his nurse or other children, that other people perceive similar phenomena; arrives at the conclusion that these other people are real; then, by a tremendous effort of imagination, *invents something which is not the phenomena* to occupy that point in space; then, reading the laws of Sir Isaac Newton, arrives at the notion of 'mass' as the occupant; and, finally, just about at the time he is leaving school, learns that his limbs—being composed of fixed quantities of Newtonian 'mass'—will accelerate in proportion to the amount of force he applies to them. This discovery, made in the nick of time, enables him to perform the motions necessary to take him to a university.

The fact that we are equipped with a special psychological apparatus for discovering the physical world, without having to call upon any sensation save that of pressure, came to me as a considerable surprise. I had imagined before that the physical universe was something which, somehow or other, we *abstracted* from such sensations as light and sound and heat and cold. But none of these is involved. Pressure is the only sensation required. Consequently, with the acceptance of P, S and T as terms for physical description, (as we have done everywhere in Part III), we have a complete physical universe running through from the remotest visible star in A_1, to the ultimate psychological observer at the unreachable end of our table.

It is interesting to observe how this direct acquaintance with the physical world, by means of the sensation of force, is related to the remainder of the sensations. You are constantly changing these other psychological phenomena. Your eyelids tire, and you let them fall. Immediately, a previous visual phenomena vanishes. You move your hand; and, forthwith, a previous unpleasant feeling of heat disappears. In such cases, you, the psychological observer, interfere. But it is important to note that you do not interfere directly with the sensation. You close your *eyelids: you* remove your *hand.* And the eyelids are not the visual phenomenon; the hand is not the sensation of heat. Here you become aware of a new class of objects, existing independently of the purely subjective sensory presentations—the colours, lights, sounds, etc. You may open and close your eyes in darkness, when there is no visual phenomenon to be observed. You may move your hand when it is touching nothing. And experiment shows that, if we classify the ordinary psychological objects as phenomena observed, we can classify this second class as observational *facilities* and observational *restrictions.* It is with this world of *facilities and restrictions* that we interfere when we change an elementary phenomenon.

We may pause here to note that one value of the physical universe seems to be that it ensures a community of experience without which we should be eternal strangers to one another.

We see, then, that the physical world constitutes a thread running straight through the hitherto separated sciences of physics and psychology.

The ultimate source of the energy transferred to the external world in the course of an experiment is the psychological observer himself. *He is* the regressive physical entity. So the question arises : How are we to bring *brain* into our table?

CHAPTER XXI

THE PLACE OF BRAIN

One method of ascertaining the connection which exists between the world of phenomenal objects and the observer's physical brain is to get hold of another fellow, poke his nervous system, and listen to what he says about it. His remarks may or may not be instructive; but, since he can talk, you will gather more information by listening than merely by watching what he *does* about it. The scientific observer, however, is not really dependent upon outside assistance, so far as regards the discovery of the mere fact that the physical correlates of a psychological phenomenon involve his nervous system.

Consider again that classical illustration of a psychological phenomenon: the globe of colour you call an 'orange'. Interpose your hand between the phenomenal object and your eyes, and the presentation vanishes. You have grounds, then, for saying that the phenomenon has a physical 'correlate' external to your eyes. But now, press with your finger on the corner of your eyeball. The phenomenon alters its shape. Further, it is possible for you to sever your own optic nerve, when the psychological object will vanish completely. You have reason, then, for asserting that the phenomenon possesses a neural correlate. But that last discovery does not permit you to assert that the phenomenon has no correlate external to the brain. A stimulation of the nerve by something external to the brain is the essential condition to the experience of what psychologists call an 'impression'. Even when you cut the nerve (an operation which is accompanied by the impression of a flash of light) the essential stimulus is from outside the organism. Phenomena which involve no such external stimulus, e.g., the memory 'image' of the orange, are of an unmistakably different character. (It may be remarked here that an 'hallucination', according to the best authorities, involves some external stimulation of the nerve endings and the illusions consist of a misinterpretation of the nature of that stimulus.)

Precisely similar considerations apply if you trepan your enemy Smith and look at his brain. Seeking for the physical correlates of your consequent visual phenomenon, by the simple method of exploration with your hand, you find that these comprise a connected chain of physical objects starting with Smith's brain and including part of your own. The method, of course, leaves you ignorant of any but the most macroscopic details of the chain, but it suffices to assure you that you—as the psychological observer of phenomenal objects at the 'now'—must place your own brain in the same world as Smith's, viz., among the A_1 physical correlates of the phenomena.

Tabulating, then, the regressive observer of impressional phenomena, and identifying him, at this stage, with C_1 we fill in his A_1 compartment as follows:

A_1
Impressional Phenomena *paralleled by* Brain affected by an external object

His B_1 compartment will contain:

B_1
Observer of nervous energy associated with Impressional Phenomena A_1 *and* Physical Interactor with Brain A_1

If he, C_1, is merely a thinker manipulating the so-called memory 'images', the A. compartment will contain:

A_1
Memory Phenomena *paralleled by* Internal activity of Brain

And B_1 will be:

B_1
Observer of Physical Mnemonic Phenomena A_1 *and* Physical Interactor with Brain A_1

What started us along the time regress, however, was the search for the source of the energy which makes its way into the object world in the course of every experiment. It might be stored in the observing instrument; but, on bringing that instrument into the picture, the time regress compels us to realise that the source of the energy which releases the stored energy in the instrument (if such there be) is still to seek. The result is the infinite regress of a source of energy. Now, we know, all of us, that the energy which initiates an experiment with an instrument comes from the experimenter's brain. And I suppose most of my readers expected (as I did myself) that brain would enter the regress as the observer C. We see now that it does nothing of the kind. The experimenter's interfering brain comes into A_1, with all the rest of the objective physical world, including the physical instrument we employ subsequently as B_1.

And this brings us back to the fact to which I drew attention at the beginning of the last chapter. In experimental physics, we take what is actually an A_1 object selected from the external world, and employ it as a means of observing some other object in that same A_1 world. We shall see, in a moment, that it is the psycho-physical observer C who makes that selection. The external instrument is the external object which affects brain, in the first of the present tabulations. But there are many such external objects and many corresponding affectations of the brain. C takes his choice. But there is a limit to what he can do along these lines. The operator of that selected instrument is the A_1 brain; and the real B_1 is the physical individual who employs this A_1 brain plus this A_1 object as a means of studying some other suspected A_1 object which may not be affecting brain at all. He himself, the physical B_1 is situated at the time 1 'now' and is travelling along the fourth dimension. You, as C, select an A_1 object as an instrument, and make it travel with him. That process is simple enough. The selection of an A_1 for use as a B_1 involves merely that you, as C, interpret it as a three-dimensional *entity* of changing character instead of as the changing contents of a mere travelling field of view passing over A_2. Actually, what the physical B_1 observes is a travelling sectional view of the *brain* A_2. That view is his A_1. You, as C, treat *that* as a three-dimensional entity which is changing its character, and so you convert it into a companion *of* B_1 travelling with him. Any external instrument which is affecting that neural companion is being treated, consequently, as a third party to the plot—it also is regarded as travelling along time 1. But the only entity which is really travelling from C's point of view is the physical B_1.

Thus, the real time regress in the world of physics is the regress of the observer who lies behind all nervous matter—a physical creature indeed, but one confined to the realms of biology. It is that creature whom we imitate when we use our clocks and measuring rods to map out an object world in terms of time. And we can carry *that* process only one stage of the regress, the stage where an instrument is treated as a B_1, and C is merely imagined.

But, this being the case, *what about the regress of h?* It cannot regress more than one term, from the object world to the B_1 instrument ! For there can be no h in the uncertainties of the remoter observer who observes psychical phenomena: he is far too coarse a creature to respond to anything so ultra-microscopic as a single photon. Obviously, then, h must be something which *we* put into the instrument when we regard the latter as an *entity* of changing character travelling along A_2 and abstracting sectional views therefrom—something which *we* insert when we treat that instrument's temporal endurance as in the fifth instead of the fourth dimension.

But that is an investigation which deserves a new chapter.

CHAPTER XXII

'h'

Let us glance back at our table of abstractions on page 75. We see that the travelling, three-dimensional B_1, consisting of energy PS, *abstracts* energy from the four-dimensional world A_2 possessing the dimensions $PS \times cT_2$. We can find no fault with that. To 'abstract' is merely to pick out a character, as a dynamometer picks out force P from momentum PT, or as a tape measure can discover lengths within the area of a tennis court.

But, in the world of physics, B_1 does not merely abstract' energy: it *subtracts* it. Energy is actually *transferred* from A_2 to B_1 in the course of an observation, and is passed from B_1 to A_2 in the course of every interference with A_2 for experimental purposes.

Now, A_2 is a four-dimensional quantity. And you cannot *subtract*, as an independently existing thing, a three-dimensional component from a four-dimensional thing. If you reduce A_2's energy component, you reduce the magnitude of A_2's content $PS \times T_1$; just as, if you reduce the length of your tennis court, you reduce its area. Now, you can take away from the area of your tennis court and add what you have gained to the area of your flower-beds. But you cannot borrow from an area and say that you have utilised the borrowed bit in increasing the length of a line. We cannot pass PS from A_2 to B_1 without robbing A_2 of a portion of $PS \times T_1$ and utilising it nowhere.

The most obvious thing to do seems to be to add a little time 1 thickness to B_1. Unfortunately, that is just what we are unable to do. For B_1 *is* moving through the four-dimensional world with the velocity c, and, according to our regressive Relativity, this velocity is as critical in four-dimensional space as it is in three-dimensional space. B_1 can have no thickness in the direction of its travel.

Very well, suppose we give up all this business of imitating the B_1

observer with instruments external to brain. All said and done, it was we who converted an A_1 mass of metal, mirrors, prisms and what-not into a B_1. We did that simply by regarding it as a three-dimensional *entity* of changing character, instead of as a travelling, sectional view of a more real entity A_2. Let us drop that interpretation, and regard the thing as an A_1. Then it will extend in time 1 as an A_2 accompanying the object A_2. We can let the real B_1 of the regressive psycho-physical experimenter serve to determine the 'now'.

That, I am afraid, will not help us. For the regress we, actually, are following *is* the regress of that psycho-physical individual. It is from him that there comes the inflow of energy to the physical world A_1. And it is the passage of energy between *his* A_2 (i.e., brain) and *his* B_1 which is, really, our difficulty. If the trouble can be got over in his case, it can be got over in the same way in the case of the instrument and the object, where both these are in the world external to brain.

But the fact that we can, if we please, re-convert our B_1 instrument into an A_1, merely by interpreting its changes in a different fashion, is of immense importance in our problem. For, when we do this, we are, as I said before, re-converting our B_2 into an A_2, and can regard this A_2 as modified by the original object A_2. Suppose we do this whenever we think of the instrument and object as interacting. We can, immediately afterwards, treat the instrument in the other fashion, i.e., regard it as a B_2 which *has* collected action from the object A_2.

Now, it is important that the reader should grasp the fact that there is no 'fake' in this purely mental operation. *it is absolutely legitimate for you to regard a three-dimensional object either as (1) an entity situated at your own travelling psychological 'now',—an entity which is changing its character,—or as (2) the view which a four-dimensional entity presents to your travelling psychological 'now'. When you are employing that object as your source of information about another object, you are regarding it as (1): when you cease to consider it as such a source of information, you are regarding it as (2). The change in your method of interpretation involves no logical error of any kind.*

The reader will find an illustration of view (2) on page 36.

But, now, consider what is the result of this change of interpretation— the result in your five-dimensional map. Your B_2 line runs no longer athwart that world in a continuous fashion like the line in FIGURE 11. It goes, instead, like this:

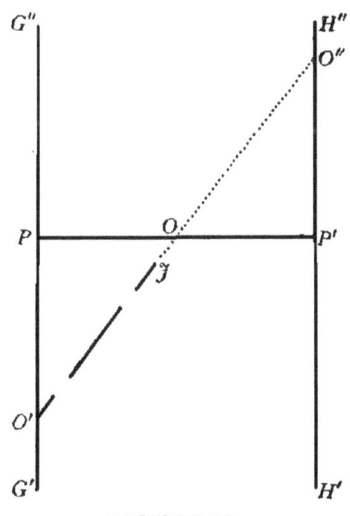

FIGURE 23.

(For simplicity the interior vertical lines of FIGURE 11 are omitted.

The breaks between O' and O show where, when PP' was passing those places, your prospective instrument (your prospective source of information) was an A_1 interacting with another A_1, i.e., was a part of the substratum with its energy changes mapped out in the fourth dimension. The places where $O'O$ is unbroken show where this entity was no longer interacting, and was being regarded as an available source of information concerning that A_1 object in the substratum with which it *had* been interacting. The dotted extension above J indicates merely your uncertainty regarding the change in the momentum and position of the prospective instrument consequent upon the last interaction with the substratum.

It will be noticed that the breaks—the discontinuities—are of different lengths. Obviously, you can start and interrupt the interactive process whenever you please.

The essential point is this : spasmodic interaction between two substratum entities would not make the endurance of either entity discontinuous. But your mental operation treats the history of your instrument as a number of discontinuous bits of an A_2 alternating with broken pieces of a B_2. Now, in the measurements of a B_2 quantity, the energy is already discrete. (A body may possess definite and limited amounts of energy.) Consequently, since both components of B_2 are discrete, B_2 itself consists of discrete portions of action of varying magnitude. Similarly, each portion of the broken A_2 line represents a discrete bit of $PS \times T_1$.

The employment of this perfectly legitimate mental device is subject, however, to certain restrictions. You must not forget c, the rate of travel of the 'now'. You must not interfere with P or S, which are unaffected by the question as to whether your instrument is a travelling B_1 or a travelling view of an A_2. So you must not lose energy PS in the course of the operation. Planck, finding himself faced with the necessity of considering action as discrete, owing to the behaviour of 'black body' radiation, found that c and the constant called the Absolute Temperature and yet another constant, Boltzmann's k, would require to be taken into account. The last two are connected with the 'entropy' of the external world, which gives the *sense* of the travel of the 'now', and so must be taken into account by ourselves. Planck did not pretend, of course, to know *why* action should present itself to us as discrete: he supposed this discontinuity to be an inexplicable attribute of the object world. But he discovered that the restrictions involved in the acceptance of these four constants—which are *our* restrictions—would limit the size of the discrete portions. They would have to consist of the atoms called h.

The fact that action, if *discontinuous,* must be also atomic is a logical consequence of the existence of the classical constants c, k and T, and of the two classical laws of energy distribution (Rayleigh's and Wien's) derived from the use of those constants. Thus, the only puzzling part of quantum theory is : *How can action be discontinuous?* It is this that serialism explains.

So, purely classical physics, when combined with serialism, suffices to account for the discovered atomicity of action.

And there's your quantum!—perfectly logical, and involving no breach of continuity in anything save the interpretations of the ultimate observer. And it is a quantum which pertains, as we had expected, to the instrument and not to the object.

CHAPTER XXIII

CHRONAXY

'Very pretty,' says the reader, (so I hope), 'but you have forgotten one thing. Your h proves that the physicist is describing his world *as if* it were being observed by an imagined serial observer. And he cannot obtain the "now" he requires for that purpose unless he himself is a real serial observer. But then, he, as this real serial observer, is confronted by the same difficulties as confront his *imagined* four-dimensional individual. He cannot pass energy in a continuous stream between his psycho-physical B_1 and his A_1. brain. He, as a four-dimensional individual, must treat the time 2 extension of his B_2 as discontinuous—must accept nothing but discrete lumps of *action* from his A_2 brain. Now, if he does that, the effect should be observable in brain *whenever he interferes* with that organism. And it should be a *large scale* effect; for he is a macroscopic individual. I cannot accept your *h* as the solution of the problem in his case. And, remember, Nature will have a say in the matter. He will find limits of some kind to the jumps of his B_1.'

And so he does.

This discovery was made by Professor L. Lapicque, and has been studied in great detail by himself, Bourguignon and Haldane. Possibly there are others who should be mentioned, for the discovery is, now, several years old.

Suppose you apply an electrical stimulus to a nerve. It will have to be a motor nerve, if you are to observe a measurable effect, but nervous matter is of the same kind everywhere, and it is with the *physical* response of the nerve that we are concerned. It is found that the intensity of the stimulus necessary to produce a response from the nerve varies inversely as the duration of that stimulus.

That means merely that the nerve responds to energy, for intensity is energy divided by time.

But it is found that there is *minimal duration* necessary to produce a response. It is an extraordinary fact that, if the duration is of less than this minimal duration, there is no response, *no matter how intense the stimulus!* Conversely, there is no response unless the stimulus has a *minimal intensity,* no matter how long the duration. That seems easier to understand. But the point is that the *minimal* intensity multiplied by the *minimal* duration constitutes the energy pertaining to an atom of action so far as the nerve in question is concerned. *[*] It is true that this atom of action is unlike the quantum, inasmuch as it is composed of an atom of energy multiplied by an atom of time; but that does not make the action other than atomic. It means merely that the character of the atom of action in the physiological world is more restricted than is the character of *h*. It is four-dimensional, but it has to possess a certain four-dimensional *shape;* whereas that shape in *h* is elastic. Again, the physiological atom of action varies with different nerves, but there is no reason, in our theory, why this should not be the case. For the ultramicroscopic world, which has to be taken into account in the ultra-microscopic experiments possible with the refined instruments of our laboratories, means nothing in the coarse reactions of living matter. The minimal intensity and minimal time can be, consequently, private idiosyncrasies of each biological structure, and even vary at different stages of that structure's life-history.

* For, since intensity and time are both minimal, we have reached a state where. energy is the smallest recordable, minimal time so that the energy also must be minimal. The action is then, minimal energy x minimal time.

So there is your discrete action in the case of the world of living tissue —in the psycho-physical experimenter's A_1!

Chronaxy in the muscles and in the sensorimotor arcs of the spinal level must be purely automatic. But that means nothing. Every physiologist knows that a flow of nervous energy which appears, at first, to be controlled becomes, with constant repetition, entirely automatic. The psycho-physical observer—observer of sensations and interactor with brain—has a physical character, and what becomes automatic in nerve or muscle should become similarly automatic in him. Since his B_1 must be the thing which makes living tissue different from dead tissue, we would expect to find it present, but habit-bound, in every tissue showing automatic chronaxy.

It should be understood quite clearly that this psycho-physical B_1 is not brain. For he can use one part of the brain and body to observe another part. When you press your finger into the corner of your eye in order to distort a visual phenomenon, you are discovering your eyeball with your finger, which observes the resistance. You can use your right hand to discover the left—and then reverse the process. In such experiments, the motor system is an A_1 object employed as a B_1; just as a camera plate is an A_1 object, being used as a source of information regarding another A_1 object.

THE SERIAL UNIVERSE

PART IV
CONCLUSION & APPENDIX

CONCLUSION

We have now completed our survey, in Part III, of the evidence afforded by the exact sciences. That evidence bears out completely the conclusions arrived at, on purely mathematical grounds, in Part II. The extensions of modern science: Relativity; Wave-particle effects; the Quantum itself: these have proved to be merely examples of the fact that a time picture is necessarily a regressive picture, and one which could not be initiated save by a regressive observer aware of a travelling 'now'. If we substitute, for the real observer 1, the instruments of our laboratory, and proceed to make a time picture, we find that we are fitting those instruments into the 'now' of the real observer 1 we had hoped to escape, so that the object world exhibits itself to those instruments as it would to him, did he possess the same accuracy of observation. And we are left, still, with the fact that the source of certain energies which make their way into the external world during an experiment, and have to be accounted for, lies at the unreachable end of the regress of the real observer.

We find that the time picture studied in Parts II and III fits perfectly the table of the self-conscious observer which we worked out in Part I, and may say, therefore, that man must be a self-conscious observer employing time as one of his terms of description because its regressive character fits his needs and gives him the only kind of picture he could regard as both rational and empirically true. But we discover a great deal more than that. We find that such an observer cannot be otherwise than immortal in his own time 2, whatever he may be in anyone else's time 2. He survives the destruction of his observer 1, and survives with the whole of his time 1 'past' experience as his four-dimensional equipment. It is unalterable, because it is fitted to the unalterable past of the objective world. This constraint—this interference with his freedom—constitutes his observation of that objective world.

Lest the reader be unduly alarmed by this picture, I may say here that there is plenty of evidence to show that observer 2 is essentially a creator of imagery—imagery which seems unreal to us now, but entirely real when we glimpse it, as we do, in our dreams. But none of this last falls within the province of the exact sciences. All that these can say is that, since man views the world in terms of time, he must be immortal in time 2. And that, I think, they may say positively.

The reader who wishes to know more about the merely psychological aspects of this four-dimensional, psycho-physical being will find a great deal on that subject in the book called *An Experiment with Time.*

And now we may attempt an answer to the question we asked ourselves in the Introduction. Is the universe rational or irrational? And the answer is : Rational in everything save the ultimate observer who makes the picture. He, with his self-consciousness and his will and his dualism of psycho-physical outlook, is irrational; but, no matter how far you may pursue him, you can never discover this. For when you reach any observer in the series, and put him into the picture, he promptly transfers the irrationality to the observer next behind him. Thus, rationality, in the philosophy of an epistemologist, lies in an infinite regress. To a metaphysician, it lies in refusing to consider any subject-object relation whatsoever. And that involves the denial of all knowledge obtained by experiment.

The reader is at perfect liberty to become a metaphysician and to say that the time picture is all wrong. But he cannot then claim that the particular metaphysical picture he may favour can be tested by experiment. Moreover, that will not enable him to escape his immortality. For when he talks about 'after' death, he is reverting to the time picture, and in that picture he is immortal.

Do we desire this immortality, now that we may feel reasonably assured that we possess it? Some of us dread it, having the false notion thereof I referred to on page 37. But all of us hate, with a hatred too deep for expression, the notion of the whole of Nature being, to Life, no more than 'an indifferently gilded execution chamber', 'replenished continually with new victims'.

But, for me, the question resolves itself very simply. There is adventure in eternal life. There is none in eternal death. And I am all for adventure.

APPENDIX

Extract from 'An Experiment with Time'

We may, conveniently, carry the analysis one stage further; but we need not trouble to repeat the arguments.

We shall discover, of course, that the time and the field and the observer, which, in stage 2, we considered as being ultimate, were not ultimate at all; and we shall come upon a larger-dimensioned lot of ultimates which, in their turn, will only retain that status until the next stage is reached. And so on to infinity.

In FIGURE 25 we exhibit three dimensions of time as the three dimensions of a solid figure seen in perspective. We have to draw imaginary boundaries to this figure in order to make the perspective clear; but, *actually, there are no such boundaries at the top or the bottom or the back or the front. The figure has fixed sides (representing birth and death* in time 1), but its extensions in the time 2 and time 3 dimensions have no limits.

Time 3 is shown as the vertical dimension of the block. In relation to this time the dimensions we call time 1 and time 2 are akin to dimensions of space.

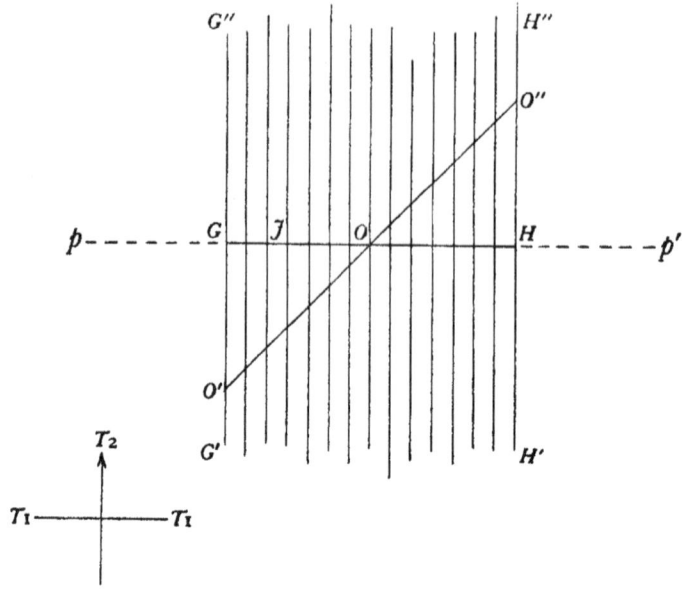

FIGURE 24.

The middle horizontal plane-section of this block-figure, the plane *G'G"H"H'*, is our instantaneous photograph Of FIGURE 24, shown in perspective. The endurances, in the new dimension of time, of the cerebral states represented by the time 2 extended lines in FIGURE 24 should be shown by extending these lines in the time 3 dimension so that they form vertical planes arranged like pieces of toast in a rack. But to fill these in would overcrowd the diagram. Our first reagent, *0'0"*, will endure (extend) in time 3 as a plane dividing the block diagonally; that is to say, the plane *ABCD*.

In the present ' condition of FIGURE 24, (shown in the middle of the block), the field of presentation *GH*—which, be it remembered, must be marked out by the intersection of some observing entity with the plane of the figure—is at the middle of the plane. In the 'past' condition Of FIGURE 24 (the plane at the bottom of the block) this field—this line of intersection—is at *DE*. In the 'future' condition Of FIGURE 24 (at the top of the block) this field is at *FB*. The intersecting entity, reagent number 2, lies, therefore, along the sloping plane *DFBE*, which plane represents its endurance.

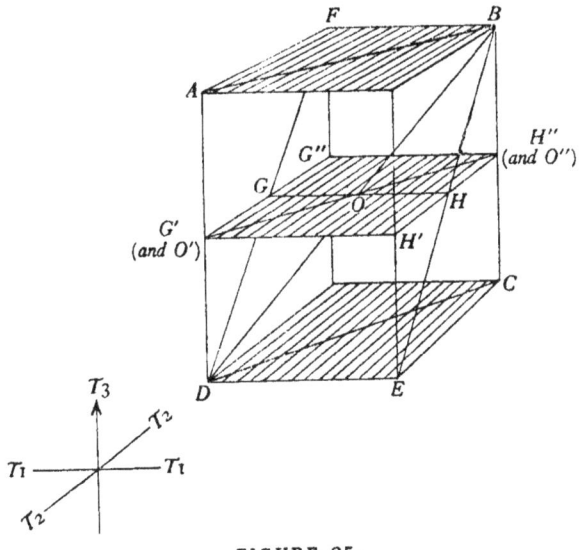

FIGURE 25.

The intersection of this plane with the plane *ABCD* is the line *DB*. The new travelling field of presentation (field 3) is the plane *G'G"H"H*. As this field 3 plane travels up the block, its line of intersection with the sloping plane *DFBE* (the line *GH*) moves over the travelling field 3 plane towards *G"H"*. That is to say, field 2 moves along time 2. The point *0* (where the three planes *ABCD*, *DFBE*, and *G'G"H"H'* intersect) moves, meanwhile, along the travelling line *GH* towards *H*. That is to say, field 1 moves along time 1.